ROAD TRAFFIC NOISE

ROAD TRAFFIC NOISE

A. ALEXANDRE and J.-Ph. BARDE

Organisation for Economic Cooperation and Development, Paris, France

C. LAMURE

Institut de Recherche des Transports, Bron, France

F. J. LANGDON

Building Research Establishment, Watford, England

A HALSTED PRESS BOOK

JOHN WILEY & SONS

NEW YORK—TORONTO

PUBLISHED IN THE U.S.A. AND CANADA BY
HALSTED PRESS
A DIVISION OF JOHN WILEY & SONS, INC., NEW YORK

Library of Congress Cataloging in Publication Data
Main entry under title:

Road traffic noise.

"A Halsted Press book."
Includes index.
1. Traffic noise.   I. Alexandre, Ariel.
TD893.6.T7R6   363.6   75–14297
ISBN 0–470–02160–8

WITH 73 ILLUSTRATIONS AND 28 TABLES

© APPLIED SCIENCE PUBLISHERS LTD 1975

Printed in Great Britain by Galliard (Printers) Ltd Great Yarmouth

PREFACE

Why a special book on road traffic noise? Simply because traffic noise is the worst offender—the most annoying source of noise and the one which affects the greatest number of people, at least in the UK and other European countries.

Twelve years have gone by since the Wilson Committee stated that 'road traffic is the predominant source of annoyance and no other source is of comparable importance', yet little seems to have changed. Or if there has been a change it has been for the worse and not for the better. Traffic noise continues to increase both in time and space. The period of quiet at night is becoming shorter and shorter, while the residential suburbs with their inadequate public transport services are suffering from increasingly heavy local traffic.

Public opinion, however, is now reacting sharply, and an additional constraint has appeared in the form of the energy shortage. Some of its implications favour reduction of noise—lower speed limits, smoother driving, greater use of public transport, and so on. Others, such as the demand for cars of smaller capacity but higher engine speed, could make the noise situation worse. In the current situation then, what should, and what can be done?

It is the purpose of this book to try to answer these questions by presenting a comprehensive assessment of the effects of traffic noise on people, and by outlining the technical, administrative and economic tools which could be employed in keeping these effects to a minimum.

The scientific data, surveys and experiments described here draw widely on research, not only from the UK but also from France, Germany and the USA, which is often less well known. Perhaps exemplifying this international approach to a common problem, authors from both sides of the Channel have come together with their differing contributions.

Ariel Alexandre, from France, is a technical administrator with ten years

v

experience of noise problems, originally with the Medical Faculty of Paris University and at present with O.E.C.D. He is also the joint author of a recent book on noise,* together with Jean-Philippe Barde, also with O.E.C.D., who is an economist involved in the problems of the environment.

Claude Lamure is a French engineer specialising in environmental problems. He is now the Director of the Transport Research Institute's Centre for the Study of Nuisance at Lyons.

Finally, John Langdon, a British social psychologist, has had a lifetime's experience of research on noise and other environmental problems and has for many years been responsible for Human Factors Research at the Building Research Establishment in England.

Each of the authors has contributed those chapters which deal with topics of primary interest to him: John Langdon, the effects of noise on man; Claude Lamure, the technical means of reducing traffic noise; Jean-Philippe Barde and Ariel Alexandre, economics, legislation and the overall planning of the book.

Naturally the authors have drawn upon the latest studies in this field and have tried to give an overall picture of the situation to date. But the scale of the effects produced by road traffic and the resulting public concern have generated a vast literature of polemic, of research and of its application, and one which continues to pour forth from the press. . . .

In this situation only a tiny fraction of this work can be cited. A full and exhaustive coverage would yield a book ten times the size of the present one—without materially adding to our useful knowledge. The authors have therefore restricted themselves to what they judge to be essential and major contributions in presenting what is believed to be the first attempt at a general survey of the problem.

The views expressed here are naturally those of the authors and not necessarily those of the institutions with which they are associated.

---

* *Le temps du bruit* (The Noise Age), Flammarion (Paris) 1973.

CONTENTS

CHAPTER 1

NOISE AND MAN

F. J. LANGDON

WHAT IS NOISE?

Before attempting to discuss traffic noise, primarily a social and community problem, it may be useful to consider briefly just what noise is and what its effects are generally thought to be.

It is not easy to define noise in a wholly satisfactory way. In acoustics and electronics it is usual to distinguish two kinds of sounds, noise and signal, the former being regarded as lacking the particular physical characteristics which define the signal. This tends to mean that noise is considered either uniform or random; or at any rate, without discernible pattern. Obviously, such a definition is merely an empirically useful one and of somewhat arbitrary character, since the noise is being defined in terms of the detector characteristics or circuitry. To take one example, road traffic noise is by no means random but is distinguished by marked spectral and temporal uniformities.

The problem of definition is not confined to the realm of physics but emerges also at a biophysical level. Thus, many definitions in this field speak of noise as sound unwanted by the hearer[1,2] and this definition cuts across the physical distinction of signal: noise. A radio turned up very loud may be conveying perfectly meaningful signals to the involuntary listener who nevertheless perceives them as noise. On the other hand, confused crowd sounds at a football match convey little or no information—they are merely the noise of the crowd—yet a football match played without them would be a sad and sorry affair. The 'information' conveyed is an essential part of the total desired experience, which is the match.

It may therefore be more profitable to consider noise not as something marked off sharply and distinctly from signal, or as a definite kind of sound, but rather as one polarity of a semantic continuum with signals at one extreme and noise at the other. In between these extremes there will be an infinite variety of conditions. We may also relate this continuum to the universe of feelings and attitudes, in which case we shall have a graded differential with pleasurable and non-pleasurable regions flanking a vague zone of indifference.

Desired sounds ←——— Indifference ———→ Undesired sounds

In such a model the central region may or may not be silence, depending purely on circumstance, for in real life we are rarely in a totally silent environment but are almost always receiving acoustic stimuli. The effects these have on us and the response we make to them will vary from time to time, and indeed from moment to moment, according to the nature of the sound, the time of day or night, our physiological and psychological state and what we are trying to do.

It is this complex variability, of the stimuli, ourselves as recipients, and the total situation which makes the problem of assessing the adverse effects of noise so difficult. Not only are there differences between individuals in what is regarded as noise and how much is acceptable, but there are differences between communities. It has also been observed (Aubree— personal communication) that such attitudes are mediated at a symbolic level, noise representing either progress, life and activity, or alternatively as a decline from truly civilised standards. Similar symbolic effects arise in relation to perceived sources of noise according to whether they are regarded positively or with disfavour.[3,4] In consequence, the same physical levels of sound will be assessed differently on noise or nuisance scales by people who do not share the same social attitudes or belong to different cultural groups.[5] We must therefore keep in mind that the extreme adverse effects represent one end of a continuum which becomes increasingly difficult to define and characterise the more we depart from its extremities.

THE EFFECTS OF NOISE

We can continue to make use of the idea of a noise–signal continuum in a very general way, without pursuing the analogy too closely, by looking at the various ways in which noise affects people. In general, we can say that as the sound pressure level increases the effects spread from attitudinal to behavioural and ultimately physiological effects. Thus at levels above

130 dB, noise can cause death, certainly intense pain and permanent damage to the hearing mechanism.[1] Its effects on sleep have been noted over a wide range from 40 to 90 dB, and on feelings and attitudes at levels well below this. For each of these situations, however, the effects may be intermingled between physiological and psychological. Moreover, it is not always easy to decide whether a physiological change or any response made by a person is to sound as noise or sound as signal, or even whether the observed effect may be interpreted as an adverse one.

In the case of traffic noise, most of the effects fall into the category of nuisance—attitudes and feelings of dislike, with only a very limited range of physiological effects—for the simple reason that traffic noise levels rarely exceed 85 dB for short periods. But before discussing these in detail it will be useful to glance briefly at more fundamental studies of noise effects generally, beginning with studies of sleep.

Effects of Noise on Sleep

In a recent review, Williams[6] points out that the results of any study are influenced by numerous factors, such as seemingly subtle differences in the acoustic stimuli used, the way these are programmed over time, the criteria adopted for measurement, and so on. These interact with human variables such as the physical and psychological state of experimental subjects, their sex, age and motivation, the conditions under which they sleep, how long the experiments last and what stages of sleep are observed. Because of these differences extreme caution is necessary in assessing experimental findings, in comparing and reconciling them with each other, and particularly in drawing conclusions from them to apply to everyday life.

Sleeping subjects cannot easily be interrogated or caused to perform tasks, at least anything more complex than pressing a button while asleep, by conditioning to an acoustic signal. We must therefore rely on the evidence from awakening from sleep, physiological indicators (such as the electroencephalogram, EEG, or electrocardiogram, ECG), or changes in behaviour and task performance following sleep. One such problem which arises in consequence is that of assessing changes in EEG as indicators of adverse effect or as having no special significance.

It has been known for some time that the intensity of neutral sounds such as bangs or clicks which will awaken a sleeper is higher than the sensation threshold of a person awake. However, a number of studies have shown that this is not due to a rise in the sensory threshold.[7-9] Central information-handling processes continue during sleep and enable the sleeper to respond to sounds such as children's cries or alarm signals. The difference between sleep and the wakened state is that in waking central brain processes—as indicated by EEG—and motor responses occur together. In sleep they form a graded series and only high intensity sounds occurring during certain phases of sleep can evoke complex motor responses.

The successive phases which follow going to sleep form an ordered series detectable by changes in the EEG. These have been labelled from I to IV according to their onset and relation to the waking state. Briefly summarised, they begin with the loss of the waking alpha-rhythm and a dreamy half-awake state (Stage I), passing into light sleep (Stage II), followed by two stages of profound sleep (Stages III and IV). It is believed that the greater part of profound sleep occurs in the first part of the night, lasting about two hours. Following this the sleeper passes into lighter sleep (Stage II) with alternation of another phase—sometimes termed Stage V—that of the rapid eye movements (REM) and together taking up the latter part of the sleep period. It is this REM state, occupying about 20% of total sleep, which Aserinsky and Kleitman[10] associated with dream activity.

Although we have a seemingly ordered process it also seems generally agreed that the pattern varies between individuals. Again, while it is more difficult to waken or obtain complex motor responses from sleepers in Stage IV sleep,[8] it appears no longer accepted that the graded series of sleep stages have graded quality. That is, it cannot be assumed that loss of sleep in Stage IV is more adverse in its effect than loss of Stage II or REM sleep.[6] Thus, Johnson[11] claims that whether a sleeper was deprived of sleep during slow wave (Stage IV) or REM sleep made no difference to his subsequent behaviour or test performance. Further, although sleepers are more easily awakened during later stages of sleep, when Stage IV tends to be replaced by REM and Stage II sleep, it is not known whether this is due to accumulated sleep or the operation of biological (circadian) rhythms.[12]

Meaningful sounds can therefore waken at low levels.[13] With suitable incentives and training neutral sounds of moderate intensity can awaken a sleeper.[14] On the other hand, there is no evidence that training or habituation enables sleepers not to respond to noise[15,16] and such findings have been generalised in the conclusion that 'the idea of adaptation (to noise during sleep) is a myth' (EPA, 1971).[87]

According to the amount of sleep accumulated or the time of night, we cannot be certain which, noise at varying levels will awaken a sleeper or induce shifts in the stage of sleep registered by EEG. It does not, however, follow that all such changes represent a loss of sleep quality.

When noise causes a loss in the amount of sleep by delayed onset or by wakening it has been shown that performance on the day following is impaired.[17,18] But we should be clear that in these cases it is not an effect of noise but of sleep deprivation consequent upon noise which produces these outcomes.

Where there is no loss in the total quantity of sleep the situation is more complicated and more controversial. The discovery that dreaming was associated with REM sleep was followed by studies in which REM sleep was deliberately curtailed.[19] Such subjects, it was claimed, showed pathological symptoms, mental confusion and withdrawn behaviour on days following and spent more time in REM sleep on subsequent nights—it was presumed 'catching up on dreaming'. Other studies reported similar behavioural effects from reduced Stage IV sleep. However, Vogel[20] in a series of studies showed that (a) reduced REM sleep does not reduce dreaming since this also occurs in non-REM sleep, and (b) neither task performance nor personality deteriorates (for a good discussion of this important issue, see Kales[21] and Johnson[11]).

A number of studies have attempted to measure the effects of night noise disturbance (other than awakening the sleeper). Thus, Le Vere *et al.*[22] claimed that nocturnal jet flyovers impaired performance on a number of mental tasks, though it is not clear whether subjects actually suffered loss of sleep by awakening. But it is clear that practice effects during the experiment reduced the data for comparison to the last two days and nights, while differences were barely significant. Another study[23] demonstrated some impairment of task performance by overnight noise which did not waken the sleeper. In this case again the majority of results failed to achieve significance level. Significant differences were demonstrated for particular age groups, mainly those aged above 50, though here the number of subjects involved was very small. On the other hand, a study of subjects disturbed but not awakened by sonic bangs over a 21-day period failed to show any significant decrement in performance of complex mental tasks.[24] Hence to quote a recent study:[25] 'it is generally conceded that the physiological and psychological significance of brief disruptions of sleep is unknown'.

It would seem that until comparatively recently most researchers accepted a sleep model which attributed graded value to the progressive stages of sleep from I to IV and went on to assume that shifts, such as from Stages IV to II or REM to NREM, as recorded by EEG, represented a loss in sleep quality. Thus, a recent review (OECD 1971)[89] draws upon such effects described by Schieber *et al.*[26,27] and concludes that noise affects 'the depth of sleep and hence its quality', despite the lack of evidence linking stages of sleep with graded quality or any proof that the recorded EEG changes could be shown to produce adverse outcomes in waking behaviour.

However, it would be unwise to conclude that noise which fails to awaken a sleeper has no adverse effect. Johnson[11] suggested that there may be a preferred pattern of rhythms dividing sleep between different stages during the course of the night, and this notion appears to have received support from a recent study.[28] Given severe enough disturbance during sleep and an appropriately sensitive task to measure its effects on the following day, performance may be impaired. In this study the authors suggest that where disturbing noise cuts across the 'rhythmicity of sleep patterns' the 'sleep cycle' is disrupted and some loss of sleep quality takes place.

Hence, while it seems difficult at present to draw clear conclusions about the effects of sounds which fail to awaken the sleeper, it is evident that noise which delays the onset of sleep or which awakens, during the night or prematurely, has measurable effects on task performance and on social well-being.

A number of controlled experiments have shown that these effects vary according to the age and sex of the sleepers. As already noted,[22] it has been claimed that 'loss of sleep quality' was greatest for older subjects, while the clearest demonstration of the effects of awakening noises on different groups of the population is given in the series of studies by Lukas et al.[29,30] Lukas has summarised these conclusions as follows: children between the ages of 5 and 8 are not at all affected by night noise during sleep. In general, older persons are more subject to disturbance than younger people, while women tend to be more sensitive to noise than men. Other studies[31] have also indicated that younger men and adolescents are likely to be more vulnerable than the early middle aged.

These conclusions are in line with findings from social surveys[3,32,33] which have shown that report of sleep disturbance and difficulty in getting to sleep correlate with measured noise levels only in the case of persons aged 45 years and above and, moreover, that the correlation is highest for women in the older age groups. In a more recent study the present author has found that difficulty in getting to sleep is related to night noise levels only for these older age groups and is highly significant for women above 50. It would also appear that use of soporifics as aids to sleep is closely related to noise level and to general dissatisfaction with noise. In general it may be said that previous failures to show any relationship between night noise levels and reported disturbance almost certainly arose from analysing the data only for the total population. The younger age groups, then, tend to mask the significant relationship shown by the elderly, particularly the elder women. This has been brought out interestingly in another way by Herridge,[34] who compared admission rates to mental hospitals in the neighbourhood of London Airport for residents in the noisiest and quietest areas. Herridge found the highest proportion of sleep disturbance and mental stress (sufficient to warrant hospital admission) from residents living nearest to flight paths, and among these the most subject to stress were women above the age of 50, particularly those living alone.

Such findings are not surprising if one compares them with the results of studies by Tune,[35] who had 240 respondents record their duration of sleep, time of retirement and rising, together with the number of spontaneous night awakenings. In normal circumstances, Tune found systematic variations in all these factors with age, sex and personality, in particular an increase in awakenings and easily disrupted sleep for elderly persons, especially women. Extroverts showed a sharp increase in sleep duration with age after 50 years as compared with introverts, mainly due to later hours of awakening. Morgan[36] also cites survey data enabling 'good' sleepers to be differentiated from 'poor' sleepers and that the two groups showed different responses to personality tests, the poor sleepers suffering more emotional difficulties and higher anxiety levels than the good sleepers. It is clear, therefore, that there exist intimate connections between sex, age and temperament on the one hand, and sleep requirements and patterns of sleeping on the other, which tend to render certain sections of the population vulnerable to disturbance of sleep.

In sum, the most clearly demonstrated effects of noise on sleep are delay in onset, involuntary awakening and disruption of the sleep pattern. Such effects have been measured fairly accurately under controlled conditions and have been confirmed by statistical and social surveys. On the other hand, more subtle but possibly equally harmful effects, such as reduced quality of sleep, have proved more elusive and resistant to precise measurement. However, even in the case of direct loss of sleep quantity no studies have yet attempted to measure the effects in such a way that the data can be used to establish criteria or standards for noise control. These factors can at present only be 'taken into account' in a general non-quantitative fashion by expert bodies, and this remains true even for evidence of graded responses to questions employed in social surveys.

The Effect of Noise on Performance of Tasks

It is perhaps natural to expect that mental tasks and work which calls for a degree of concentration will be made more difficult in noisy conditions.

And if the disturbance is severe we would go on to expect that this would lead to a decline in performance which could easily be measured. However, like many expectations derived from what at first seems self-evident common sense, carefully controlled experiments have failed to bear it out. This is not to say that noise has no effect on task performance. On the contrary, it can have a considerable effect, but whether this is an adverse or a favourable one, or whether there is any effect at all, depends on the nature of the task and the state of the performer.

An early study is illustrative in this respect. Viteles and Smith[37] subjected naval ratings to noise at 72, 80 and 90 dB in an environment at temperatures of 27, 30 and 34·5°C (effective temperature), for a period of seven weeks. In the course of 42 experimental sessions, each of 5½ hours, the subjects carried out six tasks involving a number of mental, physical and sensori-motor skills. The results showed that performance on all tasks declined systematically with rise in temperature but showed no impairment over the range of noise levels, from which the investigators concluded that noise has no effect on task performance.

Similarly, a very recent study[38] subjected college graduates to 115 dB of 'synthetic airplane' noise for 7-hour periods, alternating with 7-hour 'quiet' periods under conditions of 80 dB noise over a period of four weeks, distributing the quiet and noisy sessions in a counterbalanced design. During the sessions subjects performed nine tests requiring a high degree of mental and physical skill and in which they had reached a high and stable level of performance through many weeks of practice prior to the start of the experiment. Although the subjects suffered severe temporary hearing loss and showed many adverse metabolic symptoms as a result of exposure to these extremely high noise levels, only one task gave any indication of impaired performance (a test of serial reaction time), and this result is claimed only as 'indeterminate';[38] that is, the effect was not statistically significant. On all the other eight tasks performance remained unaffected, and from this study also the conclusion would appear to be that noise at any level has little or no effect on task performance.

Nevertheless, there are a number of reasons for not accepting these results as conclusive. In the first place it has been known for some time that, whatever the task, it is unlikely that performance will suffer at levels below 90 dB and this probably accounts for the negative result in Viteles and Smith's study,[37] while subsequent work has emphasised the importance of the type of task which may be affected. In the case of Stevens' experiment,[38] it is important to remember that the 'quiet' condition was 80 dB, so that comparison is being made over a reduced range between this level and 115 dB and the results are therefore likely to be misleading.

In a now historic series of studies, Broadbent[39,40] showed that with noise levels exceeding 90 dB performance declined according to the nature of the task. Thus 'vigilance' tasks, requiring continuous monitoring, continuous information handling and mental arithmetic, showed systematic impairment from noise. But the impairment varied, sometimes taking the form of reduced output, sometimes increased proportion of errors. Nor were the effects the same with noises of different spectra, high frequency having greater effects than low frequency sounds. Broadbent also showed that the attempt to measure changes in performance is complicated by the subject shifting his criteria for accepting or rejecting a signal during the course of monitoring experiments (for a survey of this work see Broadbent[41]).

Where the noise is of an intermittent character, as distinct from invariant continuous noise, Woodhead[42,43] found that only one of a large number of mental tasks suffered impairment from bursts of noise, even as loud as 115 dB. In a recent review, Hockey[44] points out that such an effect should be regarded rather as the product of a sharp environmental change than of noise *per se*, since similar results may be produced by periods of silence following loud noise.

In the experiments so far considered, noise appears either to have had no effect, or some adverse effect, usually an increase in the proportion of errors, on certain mental tasks. But there are circumstances in which noise has been shown actually to improve performance of the same type of vigilance task which is normally impaired by noise. Subjects who have been deprived of sleep will perform such tasks less well than normally rested subjects. But when submitted to noise at 100 dB their performance improves to a level comparable with those who have slept normally, while the fully rested subjects show impaired performance.[17,45] This result has been attributed to 'arousal', the high noise level acting to maintain the otherwise lowered level of vigilance of subjects deprived of sleep, while the impaired performance of the normally rested subjects is attributed to 'over-arousal'.

This explanation may seem at first sight somewhat unconvincing but receives support, partly from Broadbent's observation that the criteria adopted by subjects for accepting or rejecting signals in a monitoring task (and hence the number and type of errors) relate to the degree of under- or over-arousal,[46] and partly from subsequent studies.

Thus, where a task requires a subject to allocate his attention between a central task such as monitoring or tracking, and a peripheral task such as noting a warning light at the edge of the visual field, it has been shown that performance is affected by noise in a characteristic way. With loud noise the subject tends to ignore the peripheral cue and to concentrate blindly on the central task. This is analogous to a driver concentrating on his steering, clutch and gears and ignoring a red traffic light. In such a case, Hockey[47] reports an actual improvement in the central task—probably due to increased attention through failure to attend to the peripheral signals—but impairment of the total task according to the overall requirements. The effect of 'over-arousal' is therefore to concentrate the subject's attention unduly and reduce his 'cue-selectivity'. From this work and other similar studies[48,49] it would seem that the simpler and more concentrated a task the less it is affected by noise, whereas complex tasks requiring selective attention will be more adversely affected. On the other hand, it has been pointed out that a varying sound, particularly one containing information, such as that from a car radio, may actually improve task performance.[50]

So far we have looked at differences in the type and level of noise, differences in the nature of the task, and differences in the state of the performer, whether tired or rested. But it has also been suggested that differences in the psychological make-up of subjects can greatly influence the results. It is not difficult, in the light of what has been said with regard to arousal theory, to conceive that differences in the normal arousal level might be related to differences in temperament. If this were so it would not be surprising if noise were to affect task performance differently for persons of different personality and temperament, and recent studies indicate that this is the case.

In a series of experiments subjects were classified on Eysenck's personality inventory[51] as 'extrovert' or 'introvert'. It was hypothesised that the former would have a lower level of arousal, that is they required a continuous input of sensory stimuli to a greater degree than introverts. With a high noise level they would therefore suffer less impairment in task performance than would introverts, and if offered the possibility of adjusting the noise level themselves would select higher noise levels.[52,53] The results confirm the hypothesis, extroverts showing improved performance in a vigilance task as against impaired performance by introverts. In general it may be said that, with noise, extroverts tended to vary more in performance level, whereas introverts, though showing some impairment, tended to produce a more stable level of performance.

The preference of extroverts for higher noise levels was also demonstrated in a study using schoolchildren who were classified in a similar way by means of the 'junior' version of Eysenck's scale.[54] Here it was shown that the extrovert children were prepared to tolerate voluntarily a much higher level of white noise than introverts. The differences did not appear to be related in any way to intelligence levels or school attainment, though there were significant differences between the sexes, girls tolerating less noise than boys.

There is little evidence that noise impairs task performance notably more or less for different groups of the population. Sex differences appear to be small and unremarkable and data for different age groups seem remarkably hard to come by, possibly due to the cost and difficulty of obtaining adequate numbers of subjects over a wide age range for such time-consuming experiments.

While there are clearly demonstrated effects of noise on task performance it would seem that levels considerably higher than those normally encountered are required before such effects become evident. In addition, the tasks themselves require to be unusually boring, fatiguing and persisted in over long periods. It is therefore difficult to separate the effects of noise from other environmental factors, except where its effect is one which interferes with hearing and speech comprehension or produces temporary impairment of hearing (temporary threshold shift). We should therefore turn to a more general field in which the adverse effects of noise make themselves felt; that is, in ordinary social behaviour—the way people live and the way they feel about noise.

THE EFFECTS OF NOISE ON ATTITUDES AND SOCIAL BEHAVIOUR

In present-day conditions, especially in our towns and cities, the constant and universal presence of noise has powerful effects on our behaviour and way of life. Indeed, to a degree to which, until we reflect upon it, we have become so accustomed that we are hardly aware of it. It is these effects on attitudes and feelings, rather than the physical effects on sleep and task performance, which are the most widespread and with which we shall be mainly concerned in this book, since they are the main outcomes of road traffic noise.

When someone faces the possibility of living or working in a noise environment of more than about 70 dB, the outcome of such a situation is not very likely to be a change in behaviour of the kind we can measure in a laboratory. It is more probable that he will either try to find somewhere else to live or work if he can, which is of course a change in behaviour, or alternatively he will put up with the noise but be irritated by it, which we can regard as a change in attitude. In both cases, if we want to measure the relation between the amount of noise and the kind and degree of its effect we shall resort to field observations or social surveys rather than laboratory experiments.

When we perform an experiment on the direct, objective effects of noise we assume that we can simulate some situation. The laboratory represents a bedroom, a performance test some vital component of a work task, and when we apply the experimental results to life outside the laboratory we have to be extremely careful in claiming correspondences and analogies between the two. And so far as the subject himself is concerned, either he responds independently of his own volition, by staying asleep or waking up, or encouraged by suitable rewards and punishments he carries out some test to the best of his abilities.

When we come to study the effects of noise on social behaviour and on attitudes we can no longer make any of these assumptions. Simulated aircraft or traffic noise is unlikely to have the same sort of effect on a subject who has voluntarily submitted himself to it in the laboratory as it would when he suffers it willy-nilly in his own home or his workplace. Nor is it likely that we shall have much success in persuading a subject to simulate feelings of irritation or annoyance at will.

An attempt to perform such an experiment illustrates this difficulty. Mills and Robinson[55] arranged for a number of vehicles of various types to be driven past 57 male and female subjects, half seated facing and half with their backs to this traffic stream. Each vehicle was rated for noisiness on a six-point scale of nuisance. Results indicated that the distinction between the two adjacent scale points 'acceptable' and 'noisy' occurred at about 80 dBA, a result which agreed well with an earlier experiment by Robinson using the same technique. But it is some 7 dB higher than that obtained in a Swiss experiment, in 1959,[88] and considerably higher than the levels suggested by social surveys. (For a discussion of the prefixes and suffixes to the decibel unit see Chapter 2, page 37ff.)

The reasons for this are almost certainly related to the problem of motivation. There is a world of difference between assessing an imaginary nuisance while being paid to sit and listen to vehicles in an interesting experiment—and one which gives the participant the feeling that he is utilising a skill—as against answering questions put by a social survey interviewer about the nuisance suffered involuntarily from traffic noise at home.

Nevertheless, some people have the impression that because we are forced to study the attitudes and social behaviour caused by noise through observation, subjective ratings and attitude surveys the results are in some way inferior and less 'scientific' than those of a laboratory experiment. But this is not really so, for the fact that the observations of field studies are less precise than those of laboratory experiments while the conditions cannot be controlled at will, is counterbalanced by the completely natural and realistic nature of these conditions which do not have to be simulated any more than do the attitudes which respondents exhibit when questioned.

There is, of course, an important difference between trying to measure objective physiological effects and subjective attitudes but it has nothing to do with laboratory experiments as contrasted with field studies. This is simply that in the former case the subjects' response is involuntary and often unconscious, whereas in the latter it is the result of deliberative judgement. When we carry out social surveys we are therefore compelled to ask ourselves: is the respondent telling the truth and acting responsibly or judiciously; and if he is, does the attitude he exhibits or the assessment he makes correspond to the 'real' state of affairs? That is to say, he may rate the noise level as intolerable, but how much would he actually be willing to pay, what would he be prepared to sacrifice, in order to reduce it? And if the answer to this question were 'nothing', then we might need to query just how his rating of 'intolerable' was linked with the realities of daily life. And unless we were able to answer this in a satisfactory way it could be suggested that the attitudes exhibited in a social survey are not realistic.

This means that we must be able to test, first, whether respondents are telling the truth in a responsible way and, secondly, that this 'truth' corresponds in some way with the realities of social behaviour. It is not difficult to test the first of these by statistical methods which will tell us whether there is internal consistency in what a respondent says and the degree to which information from a whole population coheres. Such tests also enable us to decide whether the distributions of replies correspond with what we are led to expect from probability and sampling theory.

The second type of question is more difficult to answer. But here observation can come to the aid of social survey. We can try to see whether the way people use their homes or act at work fits in with what they say; and we may be able to obtain economic data to use as a criterion of the results of subjective studies. In the next chapter we shall discuss these problems in some detail. For the present it is sufficient to take a rapid comprehensive look at what we know about attitudes and social behaviour in relation to noise generally.

Although we all realise that noise affects social behaviour, causing people to act in ways other than those they would have chosen in a quieter environment, there are relatively few examples of attempts to measure this in a precise way. It may be that this is because the levels of noise which produce measurable effects of this sort are well above what people are ordinarily prepared to tolerate. But this fact itself suggests a method of measurement, for if the noise nuisance is severe enough people will attempt to get it reduced or eliminated and the various ways they set about doing so can be made to yield a scale of noise nuisance.

By collecting case histories of individual and collective complaints, protests and actions against noise nuisance, Stevens *et al.*[56] were able to relate an ascending series of noise levels to a hierarchy of responses, starting with individual complaints at the lowest levels and eventuating in community campaigns at the highest. This scale has subsequently been refined by providing allowances for the character, duration and time of

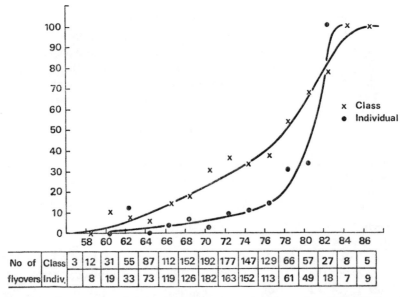

| No of | Class | 3 | 12 | 31 | 55 | 87 | 112 | 152 | 192 | 177 | 147 | 129 | 66 | 57 | 27 | 8 | 5 |
| flyovers | Indiv. | | 8 | 19 | 33 | 73 | 119 | 126 | 182 | 163 | 152 | 113 | 61 | 49 | 18 | 7 | 9 |

FIG. 1.1. The number of times the teachers paused during a flyover in 'class' and 'individual' teaching situations as a percentage of the occasions at each flyover peak level when the teacher was talking at the onset of the flyover. Ordinate: % of flyovers during which teachers paused. Abscissa: peak flyover levels inside building (dBA).

occurrence of the noise, the history of previous exposure to it, the ambient level and so on. A version developed by Kosten and Van Os[57] has been adopted by ISO. Yet, despite their accepted accuracy and fairly high prediction reliability, such procedures have severe limitations, as will become apparent when the psychological effects of noise are considered.

A further self-evident and, in principle, directly measurable effect is on the comprehension of speech. But this is a specialised topic with closest relevance to industrial and communications problems, which has been studied mainly in laboratory simulation and has generated relatively few field studies, again for the probable reason that noise producing such effects at a measurable level is more severe than community experience. Nevertheless, the effect has been used as a criterion measure in some studies of schools and of working groups. Thus Crook and Langdon[58] attempted to discover objective measures of disturbance caused to schools by jet flyovers in the vicinity of London Airport. They concluded that only the degree of speech interference, which correlated well with teachers' ratings of noise nuisance, provided any useful measure of noise effect. While unable to detect systematic increases in disruption of classroom order or changes in teaching methods with increase of peak noise, they were able to show that the total time during which lessons were interrupted bore a close relation to peak levels (Fig. 1.1).

A study of behavioural effects by Stemerding-Bartens[59] utilised speech and comprehension interference as a measure of noise effect and demonstrated subtle changes at relatively low levels by means of an ingenious technique. Group discussions were organised in 'normal' environments and noise nuisance made to occur in a seemingly accidental way. Following the discussions participants completed profiles of their fellow participants, previously unknown to them, and rated the quality of their contributions to discussion. Comparison of data from groups in quiet and noisy (70 dBA) environments enabled the investigators to measure deterioration both of speech and comprehension, and of the mutual perceptions within the groups. The masking of voice tones and simplification of language appeared to reduce interpersonal interaction and group cohesiveness. It was claimed that the effects produced experimentally could be directly related to experience in the normal work situation.

Nevertheless, in general it may be said that for the effect of noise on social behaviour to be described in a reliable way leading to quantitative predictions the levels need to be well in excess of those normally encountered. It is also a very time-consuming and laborious way of gaining information, so that it is hardly surprising that the effects of noise which have received most attention are psychological; that is, attitudes directly or indirectly expressed.

The commonest way in which this may be observed and measured is by questionnaire field studies and these have been carried out on a very wide scale in homes, offices, schools and hospitals, covering noise of many kinds whether produced by machinery, conversation, neighbours, building services such as elevators, rail and road traffic, or aircraft noise. Since this is a brief overview of noise effects, and studies of traffic noise nuisance will be discussed in some detail in the following chapter, it will be enough to look very briefly at some of these major effects without going too closely into the various quantitative measures which have emerged from their investigation.

Naturally, all the effects which have already been described and which have been studied in the laboratory, such as those on mental work or on sleep, appear once again in social surveys. And in order to assess as reliably as possible the adverse psychological and social effects of noise, and to check on the internal consistency of the overall picture formed from respondents' answers to questionnaire studies, social surveys have ranged widely over all conceivable noise effects. These can be looked at under two broad headings: first, directly reported attitudes, feelings of annoyance and dissatisfaction, or fear and anxiety. Second, the way noise is claimed to interfere with various activities or exert an unwonted influence on the way people live.

All studies of noise nuisance have involved the use of some type of rating scale, ranging from simple assessments of the degree of felt annoyance to complex 'batteries' of verbal descriptions which attempt to discriminate between the different increasing levels of disturbance. Much of the earlier work on this topic failed to draw a clear distinction between estimation of the magnitude or loudness of the sound and assessment of its adverse unwanted effect. Keighley[60] and Langdon[61] drew attention to the way in which many scales proposed for measuring adverse effects consisted of jumbled items, some related to stimulus magnitude, others to the attitudinal aspects. This is a matter of vital importance, which was stressed at the beginning of the chapter when attempting a definition of noise, for the essential thing is not the magnitude of the sound but its capacity for evoking unpleasant feelings. The sound of a dripping tap is not very loud but can generate intense annoyance if one is trying to sleep in an otherwise quiet house. Bowsher and Robinson[62] considered this specific aspect in an attempt to scale 'unpleasantness', independent of loudness or noisiness, for a range of sounds presented at varying intensity levels.

However, the recognition that sounds do differ in this way is only part of a wider problem. The effective scaling of human responses to noise requires the application of semantics in the interest of clear thinking. Failure to appreciate the importance of this requirement, perhaps because many of the earlier investigators were natural rather than social scientists, impaired many of the earlier studies. In the next chapter we shall see how this problem, which must be overcome if we are to quantify the effects of noise, has been tackled.

For the present we may note that one approach to scaling the overall effect of noise has been to construct a complete verbal scale or a 'battery' of scales,[63,64] in which each interval of a scale is represented by a word or phrase such as 'very much' or 'moderately' annoyed. A number of different scales have been employed on the ground that no single semantic dimension and no single scale is likely to yield a reliable overall assessment. As against this, other studies have attempted direct numerical assessment of annoyance, dissatisfaction or acceptability.[32,60,65,66] Nevertheless, whether verbal scales or semantic differentials linking opposed attitudes are employed they have usually been conjoined with measures of the indirect consequences of noise such as interference with daily life. And it has been argued[63] that the overall annoyance caused by noise is the resultant or sum total of all the specific activities interfered with.

This second group of effects ranges from interference with mental work, conversation, listening to the radio or trying to get to sleep, to closing windows in warm weather or dining with guests in the kitchen rather than the noisier dining room. What is important here is that noise not only affects these things increasingly with rise in level, but each is affected in different degree so that they form a hierarchy. As we might expect, difficulty in getting to sleep is rated more serious than interference with reading the newspaper. Moreover, many of these effects which form scale hierarchies are found to intercorrelate with the amount of general dissatisfaction, as suggested by Borsky.[63]

Social surveys have indicated that such ~~interrelated~~ scales form a coherent matrix representing a recognisable pattern of feeling and experience. But it is not enough to have demonstrated that such a response exists. We need to know how it is related quantitatively to the noise which evokes it.

Now the feelings and attitudes described are the same whether people are talking about life at home, in school or at work. And the way in which activities are disrupted varies only according to the job and the situation. But the noise climate which produces these effects will vary enormously.

Jet flyovers may produce peaks of noise exceeding 110 PNdB for a few seconds separated by minutes or hours of calm. Office noise may be made up of people talking and the clatter of machines and typewriters. Traffic noise may come to us as surges rising to 80 dBA every few seconds, and so on. Given these differences, is it possible to measure them all by means of one simple scale of acoustic intensity?

Studies of noise nuisance in offices[67] related annoyance to average noise level, corrected for variations in spectral frequency of the sources. On the other hand, Keighley[60,66] found that average level measured in a large number of offices carrying out a wide range of activities correlated only moderately with rated 'noisiness' and poorly with expressed dissatisfaction. But when the noise level was corrected to allow for the rate and magnitude of discrete noises rising above the average level, the resulting noise measure correlated very highly with dissatisfaction.

A similar situation may be observed in the case of nuisance from aircraft noise in which any simple measure, such as an average level over the day or even average peak level, represents but poorly the effective component responsible for the annoyance. In consequence, a number of composite measures have been developed. McKennell[64] argued that not only must the peak level be taken into account but also the number of flights per day, resulting in a combined measure—the Noise and Number Index (NNI). A similar line of reasoning was followed by Alexandre,[68] although other investigators, while recognising that aircraft noise is intense but inter-mittent, have developed alternative approaches.

As will be seen when discussing traffic noise, the effects are the same although once more the character of the noise is different. And again it has been claimed that no simple measure of intensity, such as average level, adequately characterises the amount of annoyance, so that prediction of the effect of noise cannot be made with any accuracy.[33,69] In consequence, investigators have tried to incorporate in the noise measure parameters which represent characteristics thought to be important or revealed by multiple regression analysis. Thus, traffic noise could in principle be characterised by some unit which took account of the background level and the successive peaks produced by the passage of noisy vehicles. The only problem, assuming the psychological appropriateness of such a model, is how to quantify these factors in their true relative importance.

One approach to this was foreshadowed by the Wilson Committee,[70] which developed the notion of 'noise climate': that level bounded by the background noise, i.e. the noise level exceeded for 90% of the time, and the peak noise, i.e. the level exceeded for only 10% of the time. An analogous development was to relate the annoyance expressed to the median level (in these terms the level exceeded for 50% of the time) plus some measure of the fluctuations about this, such as the standard deviation about the mean or median value. One such measure was the Traffic Noise Index (TNI) developed by Griffiths and Langdon,[65] which may be expressed in this way or by a weighted coefficient expressing the range of the noise climate related to a base level.

Alternatively, it may be accepted that while noises differ in some way, such as duration, intensity, repetition rate, and so on, so that an 'average' value cannot be used to relate the noise to human experience—since too much information is thereby lost—it may be that some measure can render them 'equivalent' in this respect. And such a notion may be glimpsed behind Keighley's 'Transient Peak Index'[66] which attempted to relate an intermittent noise climate to a steady state.

Such a unit has been developed by a number of researchers and termed, according to its origin, 'equivalent noise level'* (Leq) or 'mean energy level'.[71,72] By weighting sound levels according to their relative frequency of occurrence and expressing the logarithmic sum as a single number, a measure can be generated which allows for variations in intensity, repetition and duration so that different noises can be treated as one. With the aid of this unit investigators have studied a great variety of noise nuisance situations, ranging from the effect of road traffic in residential areas[73] and in schools,[74] and rail noise.[75]

The notion of a unit capable of scaling nuisance independent of the type and origin of the noise source is obviously a useful one and has stimulated others to widen its range of application. Thus, taking as the starting point the summation of acoustic energy represented by Leq, Robinson[76,77] suggested that the degree of fluctuation of a noise may not be adequately represented by this unit. If the fluctuation be represented by a weighting of the standard deviation about L_{50} a further unit may be generated (termed the Level of Noise Pollution: LNP) which could deal not only with any type of noise occurrence, but with different types of noise—such as traffic and aircraft—occurring simultaneously in any combination.

It is pertinent to observe, however, that none of the units so far described are very accurate, for they have succeeded in accounting for only a relatively small proportion of the variance; that is, we cannot predict an individual's response to noise with any precision.

There are many reasons for this which have little to do with noise as such

---

* Äquivalenter dauerstörpegel.

but much more to do with the respondent, either as an individual or as a member of a social group. From work already described[53,54] it might be expected that people differ considerably in their sensitivity to and tolerance of noise. More recently, Bryan and Tempest[78] have drawn attention to the wide differences in annoyance scores. Thus, McKennell[64] had noted that 15% of his sample living in 'quiet areas' below 84 PNdB around London Airport rated themselves 'very annoyed', while at the same time 32% of residents living in 'noisy' areas above a peak level of 104 PNdB were 'not bothered' by aircraft noise. It is therefore likely that typologies of sensitive and non-sensitive people yield a closer approximation to reality than the assumption of a single normally distributed population, and current work by Langdon appears to support such a classification.

Another extra-acoustical factor is the attitude to the source of noise, in a symbolic or ideological sense. It has already been noted that such differences in attitude produced marked difference in ratings for equal loudness sounds in laboratory studies,[4] and this finding is confirmed by a social survey of nuisance caused by aircraft noise.[79] Here the population residing at one half of an airport perimeter were sent 'propaganda' and public relations material on aircraft operations. Residents at the other half of the perimeter area were not contacted in any way. A subsequent social survey rated annoyance among the two populations and found that while 79% of the control population rated the noise level annoying, only 54% of the experimental population did so, although the noise level and exposure was the same for the two populations.

Again, a study of rail noise[75] revealed that less annoyance was produced by an equal amount of noise from rail trains than from road traffic and this was explained by a more favourable attitude to the former means of transport. It has also been shown that the amount of nuisance experienced is influenced by the way the respondent feels about his local neighbourhood. In the same study, as in an earlier survey of traffic noise nuisance,[32] it was found that the degree to which a person was satisfied with his locality and its facilities could influence his rating of nuisance to the extent of being equivalent to 5 dB of noise.

Another factor, discussed by Borsky[63,80] in the case of aircraft noise, is that of fear that something may go wrong; for example, the aircraft may crash on the dwelling. The more intense the noise, the more likely this fear will be experienced. This effect of 'misfeasance', to use Borsky's term, is significantly related to annoyance and may be regarded as a particular aspect of it, somewhat analogously to effects such as those of 'intrusiveness' described by Bowsher and Robinson.[62]

REVERSE EXPECTED IN USA

Finally, there is the effect of the social group or the community and the reference standards which are implicitly adopted in judging the acceptability of any situation. Thus, studies of noise nuisance among hospital inmates tended to show that patients from the higher socio-economic groups were more bothered by noise than those from lower economic levels. This was not, it seemed, due merely to expectations of greater comfort but mainly because the hospital noises were reassuring to patients belonging to the latter groups, indicating that they had not been forgotten, whereas the more prosperous patients felt less need of such 'reassurance signals'.

A study of noise nuisance related to insulation value of partition walls in three apartment blocks[81] showed that while residents in flats with partitions of 45 dB insulation value suffered more nuisance than those with 50 dB partitions, as might be expected, residents in the apartments with only 40 dB partitions complained less than those with superior insulation. The apparent discrepancy was accounted for by the fact that the third group of residents had been rehoused from very poor slum conditions and in consequence were satisfied with their new homes. Nevertheless, with time they quickly habituated to the improved standards and later studies tend to indicate that the degree of satisfaction with newly improved standards is only temporary. The long-term reference standards come to be those of the group which people judge themselves to belong to. This is in line with results obtained by comparing noise nuisance from road traffic in Stockholm and Ferrara.[5] Residents of the former city enjoyed a considerably lower level of noise but produced higher annoyance scores than inhabitants of the latter, who experienced much higher levels of noise. In this study the greatest care was exercised to ensure that the Swedish and Italian questionnaires were semantically equivalent and the results must be taken as indicative of differences in community standards and total life-styles.

Studies of noise nuisance in housing projects have indicated that external noise, particularly that generated by road traffic, has now become the dominant source of nuisance. Thus from a study of 2000 residents in houses and apartments, Chapman[82] found that at that time (1948) residents were about equally disturbed by noises within their own dwellings or caused by neighbours and noise caused by road traffic. A Dutch study carried out a little later (1955)[83] yielded very similar results—39% as against Chapman's figure of 40% complaining of annoyance from traffic and other external noises. But the London Noise Survey carried out in 1963, fifteen years after Chapman's original work, showed traffic noise to be the chief source of nuisance at 84% of the residences studied. The fact that these were located on an orthogonal grid spaced at 500 m ensured the overall representativeness of the study, so the results give the clearest indication of the rise to a dominant position of traffic noise as the major nuisance.

From all that has been said in this section, it is evident that despite the growing sophistication of social research and the proliferation of both psychological scales and noise units, the effects of noise on attitudes—self-evident to all of us, a source of frustration to citizens and concern to governments—is not easy to measure. Nor are the measures proposed as adequate as we would wish, either in their coverage of the range and type of noise situations, or in their predictive adequacy, to give that measure of control needed by planners and administrators. This is not to say that we are without predictive measures or planning instruments, but that the relation between these and a scientific explanation is still somewhat tenuous.

But the fact that we have only an imperfect understanding of the adverse effects of noise, that we cannot give a complete and quantitative account of its effects upon sleep or its capacity for provoking annoyance, does not mean that we are unable to establish standards or criteria to limit these effects. The evil effects of noise are realities which call for the measures to counteract them, of which scientific research is only a part. In an extremely sensible review of the problem, Botsford[84,85] has attacked the policy of 'attentisme', assisted by continual refinement of noise nuisance measures, which endlessly postpones effective action by fostering a 'cottage industry' of self-perpetuating noise research. There is also a social and political problem which is by no means restricted to the present day: 'Ministers being of uncertain duration and mainly concerned with safeguarding their existence before Parliament, officialdom reigns supreme and creates a device of inaction known as "The Report" which delays effective action'.[86]

As against the imperfect state of knowledge—a commodity always in short supply—stands public need and resulting public pressure which requires that nuisances be reduced by the best use of available information. In attempting, therefore, to give a sober account of what we have been able to discover of the effects of noise we must not allow our uncertainties in some areas to blur our awareness of both the need and the ability to intervene in the interests of tolerable conditions of life. And in the next chapter we shall begin to see in more detail what this amounts to.

REFERENCES

1. Burns, W. F. *Noise and Man*, Murray, London, 1968.
2. Kryter, K. D. *The Effects of Noise on Man*, Academic Press, NY, 1970.
3. Aubree, D. *Ètude de la Gêne du au Bruit de Train*, CSTB, Paris, June 1973.
4. Cederlöf, R., Jonsson, E. and Kajland, A. Annoyance reactions to noise from motor vehicles: an experimental study, *Acoustica*, **13**, 4, 1963.
5. Jonsson, E., Kajland, A., Pacagnella, B. and Sörensen, S. Annoyance reactions to traffic noise in Italy and Sweden, *Arch. Environ. Health*, **19**, 692–99, 1969.
6. Williams, H. L. 'Effects of Noise on Sleep: a Review'. Int. Congr. on Noise, Dubrovnik, May 1973.
7. Davis, H., Davis, P. A., Loomis, A. L., Harvey, E. N. and Hobart, G. Electrical reactions of the human brain to auditory stimulation during sleep, *J. Neurophysiol.*, **2**, 500–14, 1939.
8. Keefe, F. B., Johnson, L. C. and Hunter, E. J. EEG and autonomic response pattern during waking and sleep stages, *Psychophysiology*, **8**, 198–212, 1971.
9. Thiessen, G. J. *Effects of Noise during Sleep*, Welch & Welch, NY, 1970.
10. Aserinsky, E. and Kleitman, N. Regularly occurring periods of eye motility and concomitant phenomena during sleep, *Science*, **118**, 272–74, 1953.
11. Johnson, L. C. Are stages of sleep related to waking behaviour? *Amer. Scientist*, **61** (3), 326–38, 1973.
12. Rechtschaffen, A., Hauri, P. and Zeitlin, N. Auditory awakening thresholds in REM and NREM sleep stages, *Percept. Motor Skills*, **22**, 927–42, 1968.
13. Miller, J. D. 'Effects of Noise on People', US Env. Prot. Agency, NTI 0300.7, 1971.
14. Zung, W. W. K. and Wilson, W. P. Response to auditory stimulation during sleep, *Arch. gen. Psychiat.*, **4**, 548–52, 1961.
15. Friedmann, J. 'The Effects of Aircraft Noise on Sleep Physiology, as Recorded in the Home'. Int. Congr. on Noise, Dubrovnik, May 1973.
16. Johnson, L. C. and Lubin, A. The orienting reflex during waking and sleep, *Electroencephal. Clin. Neurophysiol.*, **22**, 11–21, 1967.
17. Corcoran, D. W. J. Noise and loss of sleep, *Q.J.Exp.Psychol.*, **14** (3), 178–82, 1962.
18. Wilkinson, R. T. 'Sleep deprivation: performance tests for partial and selective sleep deprivation', in *Progress in Clinical Psychology*, Grune & Stratton, NY, 1969.
19. Dement, W. Recent studies on the biological role of rapid eye movement sleep, *Amer. J. Psychiat.*, **122**, 404–8, 1965.
20. Vogel, G. W. REM deprivation, III dreaming and psychosis, *Archiv. Gen. Psychiat.*, **18** (3), 312–29, 1968.
21. Kales, A. (Ed.) *Sleep: Physiology and Pathology,* Lippincot, Philadelphia, 1969.
22. Le Vere, T. E., Bartus, W. T. and Hunt, F. D. Electroencephalographic and behavioural effects of nocturnally occurring jet aircraft sounds, *Aerospace Med.*, **43**, 384–89, 1972.
23. Roth, J., Kramer, M. and Trinder, J. 'Noise, Sleep and Post-Sleep Behaviour', 124th Annual Meeting Amer. Psychiat. Assn., Washington, 1971.

24. Chiles, W. D. and West, F. 'Residual Performance Effects of Simulated Sonic Booms Introduced During Sleep', FAA Oklahoma, Report No. AM-72-19, 1972.
25. Lukas, J. S. Awakening effects of simulated sonic booms and aircraft noise on men and women, *J. Sound & Vib.*, **25**, 479–95, 1972.
26. Schieber, J. P. and Marbach, G. 'Les Perturbations du Sommeil Nocturne par des Bruits de Decollage d'Avions', 4ᵉ Congres. Soc. Erg. Langue Française, Marseille, 1966.
27. Schieber, J. P. 'Étude Analytique en Laboratoire de l'Influence du Bruit sur le Sommeil', Rapport DGRST Centre d'Etudes Bioclimatiques CNRS, Strasbourg, 1968.
28. Herbert, M. and Wilkinson, R. T. 'The Effects of Noise-disturbed Sleep on Subsequent Performance', Int. Congr. on Noise, Dubrovnik, May 1973.
29. Lukas, J. S. and Kryter, K. D. 'Awakening effects of simulated sonic booms and subsonic aircraft noise', in *Physiological Effects of Noise* (Ed. B. L. Welch and A. S. Welch), NY, 1970.
30. Lukas, J. S. and Dobbs, M. E. 'Effects of Aircraft Noise on the Sleep of Women', Final Report, NASA, CR-2041, 1972.
31. Steinicke, G. 'Die Wirkungen von Larm auf den Schlaf des Menschen', Köln Westdeutscher Verlag, No. 416, 1957.
32. Aubree, D., Auzou, S. and Rapin, J. M. *Ètude de la Gêne du au Trafic Automobile Urbain*, CSTB, Paris, 1971.
33. McKennell, A. C. and Hunt, E. A. *Noise Annoyance in Central London*, S.S.332, HMSO, London, 1966.
34. Abey-Wickrama, I., A'brook, M. F., Gattoni, F. E. G. and Herridge, C. F. Mental hospital admissions and aircraft noise, *Lancet*, 1275–77, Dec. 13, 1969.
35. Tune, G. S. The influence of age and temperament on the adult human sleep–wakefulness pattern, *Brit. J. Psychol.*, **60**, 431–43, 1969.
36. Morgan, P. A. 'Effects of Noise upon Sleep', Inst. Sound & Vib. Res. Technical Report No. 40, University of Southampton, 1970.
37. Viteles, M. S. and Smith, K. R. An experimental investigation of the effect of change in atmospheric conditions and noise upon performance, *Trans. Amer. Soc. Heating & Ventilating Engineers*, **52** (1291), 167, 106–82, 1946.
38. Stevens, S. S. Stability of human performance under intense noise, *J. Sound & Vib.*, **21** (1), 35–56, 1972.
39. Broadbent, D. E. Noise, paced performance, and vigilance tasks, *Brit. J. Psychol.*, **44**, 295–303, 1953.
40. Broadbent, D. E. The twenty dials and twenty lights test under noise conditions, *Q. J. Exp. Psychol.*, **6**, 1–5, 1954.
41. Broadbent, D. E. *Decision and Stress*, Academic Press, London, 1971.
42. Woodhead, M. M. 'Effects of Brief Loud Noise on the Performance of a Visual Task: Bursts of One Second with a Peak Intensity of 110 dB', MRC APU Report RNP, 58/914, 1958.
43. Woodhead, M. M. Effects of brief loud noise on the performance of a visual task: an experiment with single bursts at three intensities, *J. Acoust. Soc. Amer.*, **31**, 1329–31, 1959.

44. Hockey, G. R. J. Effects of noise on human efficiency and some individual differences, *J. Sound & Vib.*, **20** (3), 299–304, 1972.
45. Corcoran, D. W. J. Personality and the inverted U relationship, *Brit. J. Psychol.*, **56**, 267–73, 1965.
46. Broadbent, D. E. *Perception and Communication*, Pergamon, London, 1958.
47. Hockey, G. R. J. The effect of loud noise on attentional selectivity, *Q.J. Exp. Psychol.*, **22**, 28–36, 1970.
48. Easterbrook, J. A. The effect of emotion on cue utilisation and the organisation of human behaviour, *Psychol. Rev.*, **66**, 183–201, 1959.
49. Hockey, G. R. J. Signal probability and spatial location as possible bases for increased selectivity in noise, *Q.J. Exp. Psychol.*, **22**, 37–42, 1970.
50. Brown, I. D. The effect of a car radio on driving in traffic, *Ergonomics*, **4**, 475, 1965.
51. Eysenck, H. J. *The Biological Basis of Behaviour*, Thomas, Springfield, Ill., 1967.
52. Davies, D. R. and Hockey, G. R. J. The effects of noise and doubling the signal frequency on individual differences in visual vigilance performance, *Brit. J. Psychol.*, **57**, 381–89, 1966.
53. Davies, D. R., Hockey, G. R. J. and Taylor, A. Varied auditory stimulation, temperamental differences and vigilance performance, *Brit. J. Psychol.*, **60**, 453–57, 1969.
54. Elliott, C. D. Noise tolerance and extraversion in children, *Brit. J. Psychol.*, **62**, 375–80, 1971.
55. Mills, C. H. G. and Robinson, D. W. The subjective rating of motor vehicle noise, *Engineer*, **211**, 1070, 1961.
56. Stevens, K. N., Rosenblith, W. A. and Bolt, R. H. 'A community's reaction to noise: can it be forecast?', *Noise Control*, **1** (1), 1955.
57. Kosten, C. W. and Van Os, G. J. 'Community Reaction Criteria for External Noises', Paper F-5, Symposium No. 12, Nat. Phys. Laboratory, England, 1961.
58. Crook, M. A. and Langdon, F. J. The effect of aircraft noise on schools in the vicinity of London Airport, *J. Sound & Vib.*, **33** (4), 1974.
59. Stemmerding-Bartens, J. 'The Effect of Noise on Groups' (in Dutch), Werkrapport. D.16, Inst. voor Gezondsheidstechniek, TNO, The Hague, 1960.
60. Keighley, E. C. The determination of acceptability criteria for office noise, *J. Sound & Vib.*, **4** (1), 73, 1966.
61. Langdon, F. J. 'Problems of Scaling Noise Nuisance', 1st Conference, CIB, Garston, 1966.
62. Bowsher, J. M. and Robinson, D. W. On scaling the unpleasantness of sounds, *Brit. J. Appl. Phys.*, **13**, 179–81, 1962.
63. Borsky, P. N. 'Community Reactions to Air Force Noise, Parts I & II', WADD Technical Report 60-689, AF 33 & 41, 1961.
64. McKennell, A. C. *Aircraft Noise Annoyance around Heathrow Airport*, S.S. 337, HMSO, London, 1963.
65. Griffiths, I. D. and Langdon, F. J. Subjective response to road traffic noise, *J. Sound & Vib.*, **8** (1), 16–32, 1968.
66. Keighley, E. C. Acceptability criteria for noise in large offices, *J. Sound & Vib.*, **11** (1), 83–93, 1970.

67. Beranek, L. L. Criteria for office quieting based on questionnaire rating studies, *J. Acoust. Soc. Amer.*, **28**, 833, 1956.
68. Alexandre, A. 'Prévision de la gêne due au bruit autour des aeroports et perspectives sur les moyens d'y remedier', Doctoral Thesis, University of Paris, April, 1970.
69. Lamure, C. and Bacelon, M. 'La Gêne due au Bruit de la Circulation Automobile', Cahiers du CSTB, No. 88, Cahier 762, 1967.
70. *Wilson Committee Report: Noise*, Command Paper 2056, HMSO, London, 1963.
71. Lang, J. 'Verkehrslärm—Messung und Darstellung', F. 35. 5ᵉ Congrés. Int. d'Acoustique, Liege, 1965.
72. Rathe, E. J. and Muheim, J. Evaluation methods for total noise exposure, *J. Sound & Vib.*, **7** (1), 108–17, 1968.
73. Fog, H. and Jonsson, E. 'Traffic Noise in Residential Areas', Report 36E, National Building Research Institute, Stockholm, 1968.
74. Bruckmayer, F. and Lang, J. Störung durch Verkehrslärm in Unterrichtsräumen, *Öst. Ing. Ztschr.*, **11** (3), 73–7, 1968.
75. Aubree, D. *Ètude de la Gêne du au Bruit de Train*, CSTB, Paris, June, 1973.
76. Robinson, D. W. 'The Concept of Noise Pollution Level', Aero Report Ac 38, National Physical Laboratory, England, 1969.
77. Robinson, D. W. 'Rating the Total Noise Environment: Ideal or Pragmatic Approach?', Int. Congr. on Noise, Dubrovnik, May 1973.
78. Bryan, M. E. and Tempest, W. Are our noise laws adequate?, *Applied Acoustics*, **6** (3), 219–32, 1973.
79. Jonsson, E. and Sörensen, S. On the relationship between annoyance reactions to external environmental factors and the attitude to the source of annoyance, *Nord. Hyg. T.*, **48**, 35–45, 1967.
80. Borsky, P. N. In *Transportation Noises*, J. D. Chalupnik (Ed.), p. 219, University of Washington, Seattle, 1970.
81. Gray, P. G., Cartwright, A. and Parkin, P. H. 'Noise in Three Groups of Flats with Different Floor Insulations', National Building Studies, Research Paper No. 27, HMSO, London, 1958.
82. Chapman, D. 'A Survey of Noise in British Homes', National Building Studies, Technical Paper No. 2, HMSO, London, 1948.
83. Bitter, C. and Van Weeren, P. 'Sound Nuisance and Sound Insulation in Blocks of Dwellings', Report No. 24, Research Institute for Public Health Engineering, Delft, TNO, 1955.
84. Botsford, J. H. 'The Weighting Game', Pres. 75th meeting Acoust. Soc. Amer., May, 1968.
85. Botsford, J. H. *J. Acoust. Soc. Amer.*, **44**, 1, 381, 1968.
86. Maurois, A. (1965) *Prometheus: La Vie de Balzac* (English edn.), p. 381. Hachette.
87. 'The Social Impact of Noise', US Environmental Protection Agency, Hearings in Atlanta, 1971.
88. 'Forschungs und Versuchsanstalt', Report No. 22637 and Appendix, Generaldirektion, PTT, Switzerland, 1959.
89. *Urban Traffic Noise*, OECD, Paris, 1971.

CHAPTER 2

THE PROBLEM OF MEASURING THE EFFECTS OF TRAFFIC NOISE

F. J. LANGDON

THE EFFECTS OF TRAFFIC NOISE STUDIED BY SOCIAL SURVEYS

Noise from road traffic may be regarded as more or less continuous sound which fluctuates from hour to hour over the day in a more or less regular fashion and from moment to moment with the passage of individual vehicles. If measured from some reception point such as the face of a building, this sound may be at a low but steady level when the road is carrying a considerable volume of traffic but is distant from the building. Alternatively, where the building is close to a lightly loaded road the average level may again be low but more variable due to peaks of noise from individual vehicles. The average level may be expected to range from 45–50 dBA at quiet urban locations to about 70–75 dBA at the noisiest spots. Here the volume of traffic is likely to be highest, the noise level will tend to be a steady one and the peak levels may exceed 80–85 dBA.

This kind of noise is obviously very different from that produced by aircraft or railways, both of which can generate higher peak levels but are intermittent in character. A more important difference is that traffic noise is generated from the entire network of roads forming the matrix of an urban environment. So, while aircraft noise nuisance is confined to the environs of a few airports or their stacking points road noise will make itself felt to a greater or lesser degree everywhere. Again, if compared with noise from factories it differs by affecting the entire length of roads, whereas industrial noise will have its worst impact at some particular spot. Finally, road traffic noise is anonymous and usually unidentifiable. It proceeds from a stream of vehicles in motion which cannot easily be intercepted—it cannot be attributed to a particular vehicle nor can the individual producers easily be controlled. Aircraft or train noise is attributable to scheduled vehicles. Office or industrial noise is traceable to a stationary source, and if there is a culprit he can be identified. Hence both dissatisfaction and specific complaints find a point of focus, often leading to remedial measures, as and when nuisance arises.

The purpose of rehearsing these facts, which are not in themselves very novel, is that taken together they outline the typical features of traffic noise which distinguish it from other noise nuisance problems. This problem is a chronic rather than an acute one, ubiquitous to urban life, mildly to intensely annoying rather than physically harmful. It will but rarely generate community actions of the sort provoked by noise around airports. This is partly because it is less intense, more continuous, less sudden in its onset. And also because its victims experience difficulty in identifying individual offenders or locating responsible agencies, nor are they easily able to formulate schemes capable of giving immediate relief. Neither shall we expect to encounter such widespread complaint of sleep disturbance, since in most locations the noise level falls as road users, like their victims, retire for the night.

We may expect homes rather than offices and factories to be most affected, since the latter have their own inbuilt noise sources and are often located away from the immediate vicinity of heavy traffic. Dwellings are also the most numerous types of building; schools and hospitals, though vulnerable to noise, do not occur in every street.

The typical effects of traffic noise will therefore be annoyance, and interference with life in and around the home. Thus it may interfere with listening to the radio, talking with friends or sitting in the garden. It may prevent people opening windows in hot weather or compel guests to be entertained in the kitchen rather than the living-room. And if severe enough it may compel residents to spend money on double glazing and air-conditioning, on tranquillisers and sedatives, or even look for a quieter place to live. And if people are under stress, whether from job or family responsibilities, or from a less than average capability of managing their own affairs, it may exacerbate this stress helping to produce neurasthenia and depression, perhaps even social and mental breakdown.

Over this dispiriting catalogue of miseries and tribulations, fusing them into a diverse but identifiable whole, hovers a general feeling of irritation, frustration and annoyance, often without precise relation to any specific interference—an unfocused malaise. It is perhaps this non-acute but universal dissatisfaction, powerless to locate the particular offender or any effective agency to which complaint may be addressed, which constitutes the 'cachet specifique' of road noise nuisance.[1]

It is noteworthy that when interviewed in social surveys residents mention traffic noise, as a topic they have felt like complaining about, more frequently than any other urban nuisance. But when asked what they have actually complained about the picture changes, for traffic noise now appears as one of the least frequently mentioned items. The explanation illustrates concretely what has been said above. All other nuisances, be they noise from neighbours, poor transport services, street lighting, sanitary or refuse collection, can be related to a responsible agency, usually local and accessible, to whom complaints may be directed, while residents know from experience that the situation may be remedied.

Traffic noise nuisance is a different matter. Unlike the other noise problems it is a general social problem open only to general solutions in the form of urban planning, traffic management and the application of general control criteria and standards. To arrive at these we have to look more closely at traffic noise and how its effects can be measured and predicted.

THE PROBLEM OF MEASURING THE EFFECTS OF TRAFFIC NOISE

In Chapter 1 we have discussed the relative merits of studying how people actually respond to noise nuisance as against sampled opinions and attitudes towards it. But from what has been said above it will be evident that the diffuse, all-pervading but non-acute nature of traffic noise makes it unlikely that it will provide much in the way of hard 'behavioural' data as a basis for control measures.

Thus if we study the various community reaction noise scales we find that around and below 60 dBA (L_{10}) there may be sporadic complaints, rising to fairly widespread complaints by 70 dBA or above. But most of the data on which these scales are based are derived from community action around airports[2] and the only extensive application of such scales outside this field is the collection of complaint data to constitute an urban industrial noise criterion.[3] But apart from the reasons for this, which relate to the nature of road traffic noise, the chief cause of the limited usefulness of such scales is that they tell us most about what happens at high noise levels—in situations where complaint reactions eventuate—but have nothing to say about the effect of noise at levels where annoyance and dissatisfaction are the principal outcomes. And this is the region which it is most necessary to explore if effective planning and administrative instruments are to be devised.[4]

A number of attempts have been made to collect data relating to the effect of noise on property values, but as will be seen in Chapter 3, this approach is open to many criticisms.

An alternative approach is to study the way in which the use of the dwelling is influenced by the level of traffic noise. But there is, unfortunately, little evidence so far from systematic studies of this sort, perhaps because life inside private homes is not very accessible to observation and we are compelled to depend upon what residents tell us. Langdon has attempted to relate the proportion of bedroom windows kept open during warm nights to prevailing noise levels. But results suggest that random variations due to the small numbers at any given site, together with the limited number of occasions when counts can be made, make it difficult to establish a base value for quiet conditions and although a relationship is apparent it is too imprecise for useful predictions.

It seems that for the present we must rely on social survey methods, which means that we are restricted to measuring attitudes and reported experience. It is this approach which has produced most of the data on which the nuisance measures we possess are based.

Before becoming involved in the details of these, however, it is essential to draw a clear distinction between two fundamentally different types of study. First of all we may want to ask a number of questions, such as how much impact traffic noise has on the population; what is its incidence at particular spots; is it increasing and, if so, how fast; can we plot its future consequences. On the other hand, we may want to know how the amount of nuisance experienced is related to the level of noise or some other physical parameter, and how well we can predict the former from the latter.

Now these are quite different types of problem and, although they may be linked in the sphere of planning and administration, so far as scientific enquiry is concerned they remain independent and generate different kinds of survey. In trying to measure *incidence* of a nuisance we are concerned primarily with its effect on a population and from this standpoint the actual noise levels are hardly relevant. All we need to know is the proportion of the population affected and how it varies from one place to another. But to give reliable answers to these questions our survey must accurately represent the population of which it is a sample. When we are trying to *relate* nuisance to noise level we are concerned not with representativeness of incidence but with having enough data to guarantee statistical significance over the whole range of incidence levels, so that we can compare differences over this range and predict differences in attitude resulting from the smallest possible differences in noise level. The amount of information obtained at any given noise level from an impact survey is governed purely by the pattern of incidence. So a 'representative' sample may yield a great deal of data at low or moderate levels, since an urban road network will always contain more quiet residential roads than noisy major thoroughfares. But there will be correspondingly less information about the effect at higher noise levels. This is not a very satisfactory collection of data from which to relate nuisance and noise level. So the choice of location for study will in each case be governed by the objectives of the survey. An attempt to attain both within a single study is likely to result in insufficient data for each noise condition, or to necessitate a heavy over-sampling through trying to satisfy this requirement while still ensuring a representative population sample.

Thus the London Noise Survey[5,22] attempted to measure the incidence of traffic noise nuisance on the metropolis using a sample based on a regularly spaced 500 m grid. This yielded an excellent picture of the impact of traffic noise on the population and a good noise map of the city. But the study was unsuccessful in trying to relate noise to nuisance, one of the reasons being that too many of the sampling points were at similar noise levels, while insufficient data were obtained from extreme conditions.

Hence most studies which have made worthwhile contributions to the development of noise nuisance measures have employed samples drawn not for their representativeness but for their ability to report effects over and throughout the possible range of incidence. But it is important to remember that the data given in such 'analytic' or relational studies do not represent actual conditions in a town or region, nor are they intended to. They merely express the relationship between a psychological attitude and its physical correlate. Moreover, this relationship itself might be purely relative since it may have no external reference but simply plots points along a continuum. There remains, therefore, a further problem of 'calibrating' the index of annoyance in order to link it with universally understood, physically observable aspects of behaviour.

THE DEVELOPMENT OF NOISE NUISANCE MEASURES

Scales of annoyance, or measures of interference with activities, must meet two basic requirements. They must be psychologically, semantically and statistically consistent. And they must exhibit significant relationships with a physical parameter. These are only minimal requirements; to be of practical utility they must be able to do a great deal more than this.

Borsky[4] has sketched out a basic model of the growth of general annoyance or dissatisfaction in which the continuum of responses along such a scale is marked by reports of interference with different activities, so forming a hierarchy. But the relation between these and the degree of annoyance is modified by other, non-acoustic variables such as the attitude of a person toward the source of noise, his feelings about the neighbourhood in which he lives, whether the noise is preventable, whether he thinks it has any effect on general health, and so on. If this is the case then the composition of dissatisfaction varies in its make-up at different points along the scale and it is evidently of vital importance to ensure that, despite this, any scale attempting to describe general annoyance or dissatisfaction should retain a consistent psychological and semantic significance throughout.

A great deal of work has gone into ensuring that this requirement is met and that annoyance scales represent the intensification of a single attitudinal dimension and do not involve excursions into other, though possibly related, areas of meaning. Thus both Borsky[7] and McKennell[8] utilised techniques devised by Guttman[9] and refined by other investigators,[10] to ensure that psychological scales met these criteria, and also maintained a constant relationship between successive scale intervals. Such scales have been successfully employed, chiefly in studies of aircraft noise nuisance.[11,12] The scales employ words or phrases as successive intervals, selected by the Guttman test procedure as representing regular increments of annoyance or dissatisfaction.

An alternative approach, already referred to,[13-15] has been to scale the intervals between bipolar opposites (i.e. good–bad), a procedure derived from Osgood et al.[16] Although it has been argued that a single scale of annoyance or dissatisfaction might have a lower reliability than 'batteries' of interrelated scales, the use of single rating scales has shown that their scores correlate highly with those of multiple-battery questions. As McKennell[6] has stated, 'a simple self-rating by informants is likely to provide a measure of annoyance that is almost as reliable and valid as one obtained from a scale carefully constructed from the answers to a series of questions'. For this purpose it would seem of little importance whether a Guttman tested verbal scale or an Osgood semantic differential scale is employed, though it has been suggested that the latter is less difficult for the respondent to complete since the intervals can be related to a state of mind not easily reducible to a word or phrase. Nevertheless, the semantic differential scale poses the problem of what precise meaning is to be attached to the numbered intervals lying between the polar extremes, namely:

(1) Verbal Annoyance Scale (after McKennell)

O	O	O	O
Not at all annoyed	A little annoyed	Moderately annoyed	Very much annoyed

(2) Semantic Differential Scale (after Keighley)

1   2   3   4   5   6   7

Completely
acceptable

Completely
unacceptable

This is not a real difficulty in practice, since both types of scale require external calibration in the form of so much annoyance, or some percentage of a population annoyed, related to some extent of activities interfered with. And this procedure provides, at the same time, an interpretation of the 'meaning' of the scale intervals.

An example of the use of scale batteries is provided by a Swedish study of traffic noise,[1] which used an 11-point index of noise nuisance exposure. Here the number of items checked off by the respondent is made to yield a percentage score of nuisance. However, as used in this study the 'battery' is rather more like a series of filters, since the reply to each successive question sets the conditions for the next. The respondent having answered a general question about annoyance finds himself specifying the conditions, frequency and duration of his sufferings. This is open to the objection that many of the later replies are forced choices conditioned by earlier replies. For to avoid appearing stupid or inconsistent the respondent builds up a picture which is more coherent than his actual feelings or experience. It is perhaps not surprising, therefore, that the results of this procedure yielded rather elegant and logically satisfying assessments of noise nuisance. Yet it provides a possible illustration of the hazardous nature of social survey methods, since it is difficult to readily conceive of tests which would determine whether the overall picture was 'true' or stereotyped.

To work efficiently a scale must yield an adequately low within-sample dispersion. In non-technical terms, persons living in similar conditions with identical noise levels—for example, the residents of a length of road—should tend to score similarly, and when they differ the distribution of their scores about the mean value should be Gaussian. But this condition is likely to be honoured more in the breach than the observance, since the degree of sensitivity to noise varies greatly between one person and another. The consequences of this, reported by McKennell,[8] have already been quoted in Chapter 1, while the values for the interquartile range taken from Griffiths and Langdon[14] and shown in Fig. 2.1 indicate a large

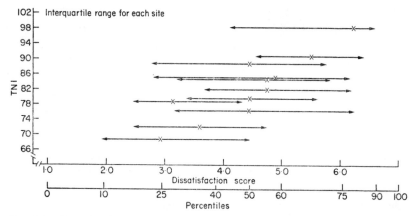

FIG. 2.1.   Plot of median dissatisfaction versus traffic noise index with the interquartile ranges shown. (After Griffiths and Langdon.[14])

measure of dispersion. In both studies only about two-thirds of the population at any given site can be located within one scale interval either side of the median value. Moreover, at many sites the distributions of annoyance scores are bimodal. Since this arises from a mixture of distinctive sub-populations, relatively little can be done to improve the efficiency of the scale. It is important to note that these circumstances set limits to the extent of intercorrelation with physical measures, no matter what parameters are adopted.

However, the efficiency of a scale is not governed solely by the dispersion of scores but is dependent also on the total score variance. So if for a given range of noise conditions median annoyance scores range over the entire length of the scale, the difference between differently circumstanced groups may be significantly greater than the differences within each group. On the other hand, even if the within-group dispersion were small the scale would not perform efficiently if the total variance is also small; that is, the shift in measured annoyance is only part of the scale even for the entire range of noise levels. Besides being efficient, a scale must also be statistically valid. That is, it is measuring what it claims to measure and not something else. This we can measure by the extent to which it correlates with other scales which measure features which can be regarded as components of the overall or general scale. So, if we question respondents about the way noise interferes with various different activities we must relate their answers to those given to the annoyance scale. From this procedure measures may be obtained, such as coefficients of covariance or partial correlation, indicating the degree to which annoyance and its presumed components increase together.

From the results of a Guttman test procedure we may be able to assume that the annoyance scale charts a regular progression from one interval to the next. But we cannot go on to assume that these distances are linear. In fact we do not know what function the scale expresses—it may be exponential or sigmoidal. In addition, because of inherent differences in individual sensitivity to noise we cannot expect normal distributions for groups of residents at each noise condition. We are therefore compelled to resort to non-parametric test procedures which do not require such assumptions of linearity or normality.

Nevertheless, because such tests are statistical procedures based upon the concept of associativity they do not tell us anything about the way the components are related to the whole or to each other, so that the underlying structure of the situation which the scale is attempting to measure is not made clear. Tests of covariance between dissatisfaction and a number of activities interfered with may suggest that some contribute less to total score variance than do others. From this we might conclude that these aspects were less important and in this, knowing nothing of the structure of the complex attitudes and experiences explored, we might be mistaken.

It is here that techniques such as factor or principal component analysis[17] can help to reveal the underlying structure. They enable reported interferences, such as with reading, conversation or sleep, to be located on dimensions within a space of cartesian coordinates. In such a space, the factors or components will be located and grouped along the dimensions according to their interrelationships. Figure 2.2 illustrates the groupings, the distance from the origin indicating the 'weight' or relative importance of the item, and the angular separation its degree of relatedness to other component items. The diagram may be read somewhat analogously to a vector diagram and affords some sort of conceptual model of the structural relationships within the total situation. Thus the illustration shows that although interference with sleep is an important component of overall annoyance it is almost completely independent of interference with day-time activities and contributes less than these to overall nuisance.

These are by no means the only methods of validation and calibration, for techniques such as the method of successive categories[17] and, more recently, Automatic Interaction Detector[18] are also in current use, but enough has been said to indicate the principles involved. The important point which must be kept in mind in the general discussion of scale construction is that almost all the techniques involved remain within the field

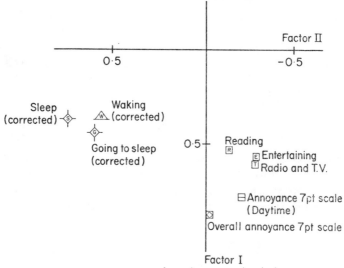

FIG. 2.2. Factorial analysis of annoyance. (After Aubree.[15])

of non-parametric statistics. This is because we cannot assume that the attitudes measured by a scale are linearly additive in a strict sense. The scale points along a psychological dimension cannot be treated as if point 2 was 'twice as far' as 1, or 3 'twice as far' as 2. No particular difficulty arises from this fact inasmuch as when the scores obtained on such scales are correlated with physical measures the coefficients merely indicate degrees of probable associativity.

But to state that a psychological X is correlated with a physical Y is not the same thing as postulating a relationship such as Ohm's law. When we do this we claim that some measurable change in resistance *produces* changes in voltage and current passed. In saying that annoyance is correlated with noise level we are claiming only that there is some probability that change of one is related to change of the other. The importance of this distinction will become apparent when we come to the problem of predicting nuisance from traffic noise.

NOISE UNITS

Chapter 1 began by drawing a distinction between sound and noise. Among the main points emphasised were (1) that when considering noise, in anything other than a purely electronic sense, we are dealing with an environmental quality, a subject–object relationship, a situation involving attitudes; (2) that the region between desired and undesired sounds does not entail silence but may contain both signals and noise. These considerations are important when we come to deal with noise nuisance, because the units we are compelled to use must of necessity be physical measures of acoustic energy, degrees of sound pressure level.

There need be no argument that physical intensity or loudness is likely to be in all cases the most important component in noise nuisance. There is therefore no question that we may as well begin by using the decibel scale, since it is in this unit that all acoustic instruments are calibrated and the great body of acoustical theory and calculation are expressed. And in general it has been agreed that for most sounds considered as noise, the decibel 'A' weighting, which is corrected over the frequency range to correspond with the response characteristic of the human ear, is a very suitable unit—certainly for the bulk of industrial, rail and road traffic noise. The effect of this weighting is to place less emphasis on sounds below 500 Hz but greater weight on mid and upper frequencies.

The only point which is essential is that whatever scales be used and whatever form decibels be expressed in, whether dBA, dBD or dB linear, whatever subscripts be appended to allow for duration, rise time, and so on, the criterion for their utility is their ability to represent human experience and response to sound as noise. This is admittedly not the sole consideration, since there are also requirements of practical utility in planning and design, but it is a very important one. This point requires stressing since much scientific ingenuity has been devoted to the elaboration of noise measures without taking cognisance of it. Botsford[19,20] has with justice pointed out that acousticians have seized upon the fact that some measure, such as dBA, fails to give weight to a particular aspect such as the duration or the rise time of the sound and have in consequence developed a more complex unit which does so, frequently one more laborious and costly to measure. Yet the whole procedure may have ignored the fact that the 'improved' unit is no more effective in representing human response than the one it seeks to displace. Thus a number of surveys, e.g. Alexandre,[21] have shown that dBA yields equally good prediction of aircraft noise nuisance, when embodied in a formula such as NNI (Noise and Number Index), as the more complex PNdB (Perceived Noise Decibel). And when we come to consider how little of the total variance may be attributed to acoustic factors this is hardly surprising.

The requirements for a noise nuisance unit may therefore be set out as follows:

1. It should give the most accurate representation of human response to the noise conditions.
2. It should be capable of being fitted within the corpus of acoustic data for purposes of calculating propagation, etc.
3. It should in principle be simple to measure without the necessity for expensive computer installations; that is, without complications such as real time analysis or third-octave band frequency analysis.
4. It should correlate well with other physical parameters; in the case of road traffic, with volume measured in vehicles per hour, etc.
5. It should be capable of being understood and applied by administrators, planners, engineers, etc. concerned with implementing environmental standards.
6. If applied to road traffic it might be advantageous if it were, in principle, applicable to situations where noise resulted from other sources as well.

With these objectives in mind researchers have employed a great variety of measures, some of which were briefly described in the previous chapter. One of the reasons for the proliferation of these has been the relatively low predictive efficiency of the indices derived from social surveys, rather than the desire of each investigator to invent a novel unit.

Thus the study of traffic noise in London carried out by McKennell and Hunt[22] failed to obtain any significant correlation between annoyance and noise level measured in dBA. Here the limited range of noise levels, the poor distribution of response to the annoyance scale, the lack of precise relationship between the resident's home and the actual point measured, and the low reported proportion of residents seriously annoyed made it unlikely that any relationship would be revealed whatever noise measures were employed, in this case the 10% and 90% levels.

A French study of traffic noise nuisance from motorways[23] employed median noise level (the level exceeded for 50% of the time) in dBA. It demonstrated that noise level correlated well with a composite index of annoyance. However, the value of 0·6 obtained for group scores is not high enough to enable the effect of reducing the noise level at any point to be predicted in terms of nuisance. Moreover, the distribution of annoyance scores was not Gaussian and the noise levels were not measured at the actual point of reception but extrapolated from roadside measurements.

A Swedish team[1] carried out a similar study in Stockholm and employed a variety of acoustic measures in conjunction with the composite nuisance index already described. The survey yielded highly significant correlations between index scores and a number of acoustic parameters, including those of peak and average level, together with a measure of energy level, Leq. Noise levels expressed in this unit are average levels in dBA, but they are averages of the fluctuating energy content of the sound wave as opposed to the averages of fluctuating dB levels, expressed by L_{50}. Thus a noise level expressed in Leq equates a fluctuating noise to the level of a steady state noise having the same energy content. By thus placing greater emphasis on the peak values of fluctuations Leq is claimed to represent more closely human perception of the noise. The values of r obtained by the Swedish study with L_{10}, L_{50} and Leq were

$$L_{10}, r = 0\cdot88; \quad L_{50}, r = 0\cdot82; \quad \text{Leq}, r = 0\cdot96$$

There is thus an apparent superiority of Leq over the other measures, although a study of statistical tables shows that the difference between the values obtained for L_{10} and Leq is not significant, and no data are given for correlation with individual responses so that a more critical evaluation cannot be made. If a pattern of distribution for individual annoyance scores similar to other noise surveys is assumed, the total variance accounted for by measured noise would amount to about 13–14%.

A similar study reported for Vienna[24] also employed Leq as a measure, though introducing small variations into the method of calculation without significantly affecting the results, which appear to be expressed in somewhat crude terms so that the evidence they provide does not significantly add to the picture.

In 1968, Griffiths and Langdon[14] carried out a similar exercise in the Greater London area. Using a seven-point semantic differential scale of dissatisfaction, with traffic noise measured over 24 hours at the facade of a dwelling, and utilising measures of background, mean and peak noise level the results given in Table 2.1 were obtained. Of these, only the correlation with L_{10} was significant, though again too low to enable noise control predictions to be made. The authors, while noting the wide dispersion of dissatisfaction scores and the departures from normal distribution, also observed that the median group scores for each site were a poor fit to the regression line, but that the fit was best for L_{10} and worst for L_{90}. They therefore sought to reduce the dispersion by relating L_{10} to the background level over the 24 hours.

TABLE 2.1

MEAN AND PEAK NOISE LEVELS MEASURED OVER 24 h AT THE
FACADE OF A DWELLING

Noise level in dBA (24 h)	Correlation coefficient with median group score
L_{10}	0·6
L_{50}	0·45
L_{90}	0·26

From a series of multiple-regression equations a formula, $L_{10} - 0·75L_{90}$, was derived representing the best overall weighting. This yielded a correlation of 0·88 between median group scores and the proposed noise unit, termed the Traffic Noise Index (TNI). This highly significant value enabled the result of reducing noise at a given location to be predicted in terms of diminished nuisance, so providing a correlative scale upon which environmental standards might be based.

However, it is also worth noting that the correlation between noise level measured in TNI and individual scores was only 0·29, so that the noise measure accounts only for some 10 % of total variance.

More recently, Aubree[15] has carried out a study of traffic noise in Greater Paris, using a semantic differential scale to measure dissatisfaction, identical in construction to that used by Griffiths and Langdon. Results reported for group median and individual scores, correlated with median noise level (L_{50}), were $r = 0·97$ (group) and 0·37 (individual). No significant difference was found between these values and correlations obtained by using L_{10} or Leq.

It would, at first sight, appear that the noise measures are accounting for a greater proportion of total variance than was explained in the other studies. However, this is not really the case since the final result is the outcome of 'correcting' the scale score for the degree of exposure, in effect, the proportion of rooms in the dwelling exposed to street noise, and also incorporating a non-acoustic variable. This is the judged quality of the local environment rated on a ten-item index. When these are eliminated the value of the correlation for individual scores falls to 0·32, not significantly different from other studies.

Finally it needs to be borne in mind that the effect of increasing the number of variables involved in the multiple correlation and regression procedure, whether acoustical, as in TNI which utilises L_{10} and L_{90}, or social, such as the terms for exposure and environmental quality, is to reduce the significance level of the correlation coefficient, since the extra information used detracts from the numbers of degrees of freedom. So, for example, while a value for 'r' of 0·8 may be significant to the 0·001 level of confidence, where a single acoustic measure such as L_{10} or Leq is used, the introduction of TNI, using information from L_{10} and L_{90}, reduces the significance level of a similar result to 0·01.

Summarising, it would seem that the British, French and Swedish surveys all sought to raise the accuracy of noise nuisance prediction, beyond the mere demonstration of a trend relationship, to the level at which it is possible to gauge the effect of a change in noise level equivalent to about 5 dBA, the minimum necessary if any noise control policy is to be implemented. In an attempt to achieve this a variety of noise parameters have been investigated, mainly with the aid of multiple-regression analysis on a trial-and-error basis, guided by some consideration of what acoustic conditions might be expected to influence the experience of annoyance.

Thus, although the mean level fluctuates over time with rise or fall in traffic volume, it might well be that short-period fluctuations occur with the passage of particularly noisy vehicles. It could therefore be argued that a measure such as mean energy level (Leq) would be particularly suited to take such effects into account, since it contains a term which is influenced by fluctuations in noise level. However, Griffiths[26] and Robinson[27] both claimed that the structure of the equation from which this measure is derived could not yield sufficient variation to give adequate weighting to such effects. Robinson proposed that the fluctuation term be increased by adding a weighted coefficient of the standard deviation of the mean level (as proposed: Leq + 2·56σ).

An alternative approach could be advocated by assuming that short-term fluctuations are taken account of by L_{10}, as this is close to the peak level and is therefore most sensitive to surges produced by noisy vehicles. At the same time it could be argued that L_{10} is not such a good discriminator of traffic conditions, since the value of L_{10} varies less between quiet and noisy locations, or at any location over 24 hours, than do L_{50} and L_{90}.

Finally, there is the consideration that the peak level, fluctuating momentarily with the passage of individual vehicles, is varyingly perceived according to the extent that it is masked by the background level emanating from the traffic stream and the adjacent road network. This effect will also depend on the distance of the dwelling from the traffic stream. For the peak level produced by individual vehicles decays over distance at double the rate of the background level from the traffic stream (see Chapter 5).

All these considerations prompted continued search for an effective measure of noise nuisance. As can be seen from the results given above, the relatively small proportion of variance accounted for in all the surveys gave a powerful incentive to further work, since without improvement in predictive accuracy little guidance could be given in establishing standards or carrying out remedial measures. Yet, in practice, not a great deal has been achieved, since the improvements in accuracy resulting from the use of L_{10}, L_{50}, Leq or TNI are all marginal, the total variance accounted for by the noise measure remaining small. In other words, the greatest part of the difference between individual annoyance scores does not appear to be traceable to differences in traffic noise level. Of course, it is easy to see by hindsight that this is bound to be so if the individual annoyance scores exhibit wide and non-Gaussian distributions. But it does not appear to have been obvious at the time when each survey was made, and the search for a noise unit did appear to be guided by what was in principle the correct criterion, fidelity to human response. However, to pursue this principle consistently it is first necessary to ascertain the causes of individual differences before being able to distinguish effectively between different noise measures, and this we must postpone until the next section.

Lamure and Bacelon[23] showed that measured noise level was highly correlated with traffic volume, a fact already demonstrated.[28] It was also shown that annoyance could be correlated with traffic flow equally well as with measured noise. Griffiths and Langdon[14] also demonstrated a similar result but were unable to provide a nuisance prediction based on traffic flow, since their data were derived only from 24-hour average flows rather than from synchronously recorded data.

However, the more recent study by Aubree[15] employed a more precise technique which enabled useful predictions to be made, at least in principle. For this purpose, hourly noise levels were summarised by statistical counters, photographed throughout the 48 hours of continuous noise measurement and the hourly levels correlated with traffic flows counted simultaneously by hydraulic road strips. Correlation between vehicle flow and L_{50} noise levels for any given site was extremely high and if street width, the main source of variation between different sites, was taken into account the relationship could be generalised for all locations ($r = 0.97$).

It would seem evident that if traffic noise can be expressed without loss of accuracy in terms of traffic flow, then noise nuisance would in principle be predictable directly from non-acoustic data. However, the French study did not go on to make this final correlation, although later current work suggests that the procedure is a perfectly viable one. This does not, of course, mean that acoustic measurements and units can be dispensed with, for these are essential if remedial measures such as sound insulation or acoustic barriers are contemplated. But it does suggest that traffic measurement offers a useful and inexpensive guide to early prediction of potential nuisance.

The earlier studies relating traffic flow to noise level contain important reservations as regards the nature and composition of the traffic. The work of Lamure and Bacelon[23] related to traffic on motorways enjoying continuous circulation, while that of Griffiths and Langdon[14] was based on a carefully selected sample of roads exhibiting similar characteristics. The traffic pattern was therefore extremely regular and very different from that typical of urban areas where parked vehicles, light-controlled intersections, pedestrian movements and frequent congestion tend to disrupt any noise–flow relationship. Secondly, the bulk of data from which these relationships have been derived are for flows in which the proportion of heavy vehicles has rarely exceeded about 20 %, while the size and power of these were smaller than that of typical present-day heavy vehicles.

There appears to exist at the moment no quantitative model for the relationship between noise level and traffic flow for congested conditions and with high proportions of heavy vehicles, although a number of formulae have been tentatively suggested.[29] These all contain coefficients for the proportions of heavy vehicles, a quantity which investigations relying on hydraulic road strips cannot provide since the counters do not yield separate totals. Work by Langdon, now in progress, suggests that with congested conditions the proportion of heavy vehicles may be a more important determinant of the noise level, and indeed of noise nuisance, than absolute volume of traffic.

Although considerations such as these may not be of immediate practical utility in the establishment of noise nuisance units, their importance cannot be overestimated. For at present there exists no general theoretical model which embodies all the factors which materially contribute to noise nuisance. Rather has it been assumed that some acoustic measure of sound intensity would suffice unaided. And while dispersion of nuisance scores remained high, and in consequence the variance accounted for remained low, there could be little apparent utility in attempting to develop such an overall model. But this situation appears to be changing and the general outlines of a systematic account of the adverse effects of road traffic noise can perhaps begin to be drawn.

A final aspect which needs to be considered in connection with any noise measure is the actual degree of physical exposure to road noise for the resident of any dwelling, since this is likely to affect the degree of nuisance experienced. The exposure will be governed by two factors: the structure and form of the dwelling and the way it is used. Of these, we can only consider the first for inclusion in a noise unit, since the second will depend upon social and cultural differences.

If a dwelling is built in such a way that all the rooms of an apartment face the street, as with some apartment blocks where a spine corridor runs through the building, residents have no option but to put up with traffic noise, night or day. In other types of building, some rooms may face the road, while others are shielded from noise by these front rooms. Aubree[15] attempted to estimate this 'exposure factor' by taking the proportion of rooms in each dwelling which faced on the street and entering this score as a physical factor mediating noise nuisance. The procedure had the effect of increasing the predictive accuracy, and although the gain in correlative value is not stated explicitly the relative 'weighting' for exposure indicates that it may be equivalent to a difference of ± 5 dBA of road noise. This corresponds to the results obtained by Lamure and Bacelon,[23] who found that for a dwelling with each room facing the road the annoyance was the same as for a dwelling where half the rooms faced the road but with noise levels 3 dBA higher.

Another approach now current is to determine the principal room by reference to the location of the television set, thereby assuming that this indicates that it is the main room in which people pass their leisure hours. If either the proportion of rooms, or the location of the principal one, can be ascertained a distinction has been made which could be expected to improve prediction of noise nuisance. Nevertheless, while contributing to the precision attained in a research project, the 'exposure' factor is unlikely to be of much practical use in application of the results. For unlike the noise level or the traffic flow it cannot be easily observed or measured from outside the dwelling, though it may sometimes be possible to estimate it from building plans.

This virtually completes the list of what might be regarded as possible physical variables which may either supplement or replace acoustic parameters. But before these can be considered in relation to nuisance scores in a general regression equation, there remain a number of intervening variables the effects of which are likely to influence considerably the overall result.

INTERVENING VARIABLES INFLUENCING NOISE NUISANCE

Although a general scale of annoyance yields a score which is in each case the sum total or the integral of the nuisance experienced, so that it may be argued that all the circumstances which contribute to nuisance have already been 'taken into account', it would be a mistake to rely solely on this. For, in the first place, annoyance scores at one location will be governed by local circumstances only some of which are directly expressed by the noise level, so that there will remain differences which are not traceable to this. Secondly, there will be differences of personality or of circumstances between individuals resident at any one location. These will influence individual scores of annoyance, so that for a given noise and environmental condition the dispersion of these will be greater than would otherwise have been the case.

Taking first those factors which may be regarded as characteristic of the whole location or site, we may note that the general environmental quality of the neighbourhood is of some importance. Aubree[15] studied this aspect in some detail, obtaining from respondents an estimate of neighbourhood quality by summing replies to a ten-item schedule. The schedule covered a range of neighbourhood amenities such as transport, schools, medical services, distance to work, provision of shopping and places of entertainment. It was presumed that the ten items offered the basis for a rough assessment of the environment, and that a resident who declared himself satisfied with six or more of the items was likely to be generally satisfied with the neighbourhood. No attempt was made to weight the items, respondents being given a score of 1 or 0 according to the number of items checked positively, irrespective of their nature. Nor was there any attempt to measure the relevance or completeness of the list or test whether the score was related to a direct assessment of neighbourhood quality.

Nevertheless, when taken in conjunction with other questions, including the dissatisfaction scale, the effect was to improve correlation with noise level to a significant degree, the value rising from 0·32 to 0·37, a gain of some 4% in variance accounted for. The predictive equation can be rearranged so that satisfaction with neighbourhood is made to yield a difference in noise level rather than a difference in predicted nuisance. If this is done, the result appears to show that such satisfaction is equivalent to a reduction of noise level by some 5 dBA.

An interesting corollary of the reported result is that the degree of satisfaction with the neighbourhood tends to be linked with noise level, one rising with the other. This seemingly paradoxical result is due to the fact that the best liked neighbourhoods have the highest provision of social amenities and tend also to have the highest volume of traffic. Hence the annoyance generated by traffic noise is in practice to some extent offset by increased satisfaction with the locality.

Now the essential ideas explored by such questions in a social survey are of the first importance. But if they are to be of practical value two conditions must be satisfied. First, we must be certain that the environmental aspects used to modify the predicted annoyance from traffic noise really are those on which environmental evaluation rests. And second, that the evaluation can be made independently of social survey questions; that it can be measured as a physical variable.

As regards the first point, the conclusions reported above appear to be somewhat modified by the results of a more recent study dealing with annoyance caused by rail noise.[30] In this study, a similar list of items was employed but submitted to a principal components analysis and intercorrelated with measured satisfaction with neighbourhood. The result of this procedure indicated that almost all the variance was accounted for by items related to the appearance and social tone of the neighbourhood. A similar analysis now being made by Langdon, using data from the Greater London area, has produced an identical result. If this is the case it would suggest that a large part of the covariance with noise level found in the earlier 1971 study was not in fact due to difference in neighbourhood quality, but merely to the inclusion of items which, being trip generators—such as shops and places of entertainment—automatically co-vary with noise level and are to this extent capable of being regarded as alternative measures of noise level. It would therefore seem essential to ascertain as carefully as possible the true content of environmental quality evaluation and to give such items their true weighting, if the effect of the general environment is to be correctly estimated.

This, however, brings us to the second requirement, for if such allowance is to be made for the environmental effect it is also necessary for this to be done by objective means. If not, its practical utility is limited to producing a more satisfactory result in a social survey by increasing the variance accounted for. Lacking some physical measure of environmental quality the prediction of probable nuisance at any given location cannot take this into account without a social survey being mounted.

This raises controversial issues going far beyond the scope of this book and entering the general field of environmental planning. In a recent study, Troy[31] has shown that not only did assessments of environmental quality made by residents differ from those of 'experts' touring the locality and endeavouring to make external, physical appraisals, but the contributing items and their 'factor structure' also differed. It therefore seems that external, physical assessment produces a picture of a neighbourhood different from that produced by residents and, if this is the case, it would appear to jeopardise the possibility of 'objective' environmental measurement of the kind which could be quickly and easily made as an additional aid to prediction of noise nuisance at a particular locality. But the requirement is clearly an important one and calls for further study, which it will no doubt receive.

The second variable which can modify the effect of nuisance, and which has already been mentioned in passing, is the exposure to noise other than that governed by physical factors such as the design of the dwelling. This covers aspects such as the life-style and activities of residents, whether they spend much or little time at home and what they do there.

The work of Aubree[15] and current studies by Langdon have brought out important features, such as the time of rising and going to bed, and these can be generalised for both Paris and London. Figure 2.3 shows that these hours are related directly to the distance from the city centres,

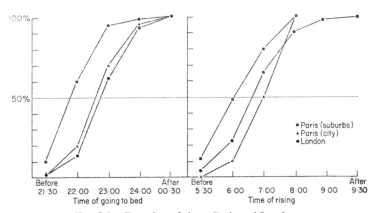

FIG. 2.3. Duration of sleep: Paris and London.

suburbanites rising and going to bed about an hour earlier than dwellers at the centre, a difference accounted for by the length of the journey to and from work. This is an important aspect to consider when looking at the level of noise for particular parts of the day and night, and especially in deciding when the night hours, in terms of acceptable noise levels for sleep, begin and end.

However, apart from this there are many other ways in which people's habits vary. At any given location a number of dwellings receive a similar 'dose' of traffic noise. But each resident or family will have a different pattern of life and activity, rendering them more or less vulnerable to its effects. Thus some will be out all day at work, others, particularly house-wives, home all or part of the day. Some people will spend much of their leisure time at home, others relatively little. Again, some will carry out study and mental tasks at home, such as making up accounts, some will attempt to play musical instruments or listen to hi-fi, while others will spend time in occupations such as woodwork, metalwork, decorating or gardening, activities relatively invulnerable to noise.

These differences between people are in no way remarkable and would hardly warrant mention but for the circumstance that they are going on at one particular location in adjacent homes. Moreover, they can be related to systematic differences in attitude to traffic noise. Thus to take but one example, Aubree[15] showed that there is a significant difference in annoy-ance score for those who spend less time at home, such respondents being more bothered than those who are at home all day.

The effect of differences in life patterns is therefore to increase the dispersion of annoyance scores for any location or site. And as these differences can only be discovered through actual survey interviews, they remain as irreducible sources of unexplained variance, setting a limit to predictive accuracy. This applies with even more force to what is probably the largest source of variance; namely, individual differences in sensitivity to noise.

Differences in life-style and resultant vulnerability to traffic noise cannot be allowed for when applying survey results to predicting likely nuisance, although they can be ascertained fairly easily in the course of surveys merely by asking respondents questions about how they spend their time. But differences in noise sensitivity are not so easily measured, since to do so requires a valid test of sensitivity.

Griffiths and Langdon[14] attempted to apply such a test but rejected it as inefficient, since its scores correlated significantly with measured noise levels, which they should not do if the test is to discriminate sensitive from insensitive persons. However, further work by Langdon suggests that a simple self-rating procedure satisfies this requirement by producing non-significant correlations; that is, the test results are not significantly contaminated by ambient noise levels. The result of separating the population into such groupings is to substantiate the observations originally made by McKennell[8] and reported here. As might have been expected, persons sensitive to noise show little variance on annoyance scales, scoring high for all noise conditions. Those identified as 'non-sensitive' by the test show a

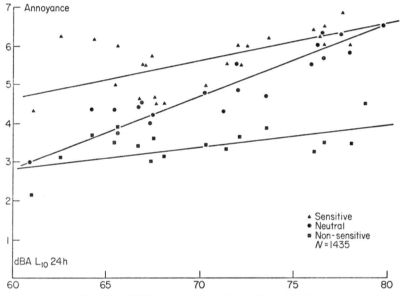

FIG. 2.4.   Noise annoyance and sensitivity to noise.

similar pattern of response, in this case at low levels of annoyance. Only those identified as 'neutral', claiming to be neither particularly sensitive nor non-sensitive to noise, show a significant change of annoyance score with change of noise level (Fig. 2.4). It is interesting to speculate on the possibility that these three regression lines correspond to different sections of an overall sigmoid curve, analogous to those employed in *Probit Analysis*,[32] in which case the three sub-populations would all show covariance if the range of noise conditions was greater than those actually encountered, as illustrated schematically in Fig. 2.5.

At present no reliable test for noise sensitivity has appeared in the published literature and, as already pointed out, even if one were to appear its utility would be confined to accounting for the variance exhibited in the research study. Bryan[33] has commented on the undesirable consequences of the apparent inability to take this factor into account in setting noise standards and this aspect will be discussed in the final section of this chapter.

Nothing has so far been said of the effect of social variables such as economic class, or factors such as age and sex. This is because surveys have repeatedly found little difference in annoyance levels with respect to these. In general, it may be said that the population as a whole appears to be uniformly affected by noise. It is possible that this finding is due to the

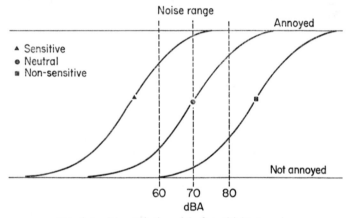

FIG. 2.5.   Theoretical model of sensitivity to noise.

relative imprecision of present methods for measuring annoyance. Since the largest part of the total variance remains unexplained there is plenty of room for significant but small variations to occur, their effects remaining indiscernible, whilst grosser effects such as those of life-style and sensitivity have not been isolated and eliminated. This is quite conceivable since Aubree[15] and others have, as already mentioned, demonstrated significant differences in getting to sleep under noisy conditions for particular sex and age groups of the population. Moreover, differences between social classes will tend to be masked by the greater capacity of those belonging to more privileged classes to obtain acoustic comfort and the tendency among underprivileged groups to accept generally lower standards of environmental amenity.

Such differences are indicated by the comparison of annoyance levels between Stockholm and Ferrara,[34] two cities with similar traffic structures but different life-styles and standards of social amenity. Here the investigators found that, although noise levels in Stockholm were between 5 and 8 dBA lower than those in Ferrara, 61% of respondents reported disturbance by traffic noise in Stockholm as compared with 49% in Ferrara (differences between the populations for the category 'greatly disturbed' are even wider at 45% and 21%, respectively), while 12% had tried to get disturbance reduced in the former city, as against only 4% in the latter.

More recently, Lawson and Walters[35] have reported on the effects of raising the traffic noise level by opening a new motorway close to apartment blocks occupied mostly by working class tenants. Two surveys were made, one before the opening of the new road, the second a year after the event. At apartments closest to the motorway the 24-hour L_{10} level rose between 9 and 15 dBA above the original levels. Although community protest was vociferous for a short time after the opening of the motorway, a year later annoyance scores had not shifted significantly from their previous levels, so that the increase in noise had been 'absorbed' by the residents. The authors comment on the remarkable capacity of the largely working class population to adjust to worsened environmental conditions. If this is so, such a tendency, allied to the greater capacity of more privileged groups to obtain locations regarded as acceptable, would, as pointed out, tend to mask differences in degree of annoyance between social classes and render them difficult to identify within the population studied by a single survey.

There remains one further factor which can influence significantly the attitude to traffic noise and this is the way in which different forms of transport are perceived. In a general discussion of the problem, Franken and Jones[36] reviewed a number of the factors which have here been dealt with in detail, in particular drawing attention to the fact that community response to noise from different transport media took different forms and that, in the case of ground transportation noise as contrasted with aircraft noise, non-physical factors other than the noise level itself played a considerable part in modifying dissatisfaction.

The experimental study referred to in Chapter 1[37] indicated that the degree of disturbance produced under laboratory conditions by noise at equal levels of intensity varied according to the type of vehicle associated with the noise. More recently, Aubree[30] in a study of annoyance caused by rail noise, has shown that road noise at a level of 65 dBA (L_{50}) generates the same degree of annoyance as rail noise at a level calculated as equivalent to 70 dBA (rail noise was measured in Leq). They attribute this difference to the favourable attitude, revealed by questions dealing with convenience, cost and social utility, to railways as compared with road traffic as transport media. These are therefore important intervening variables which need to be taken into account, in this case having the advantage that once having been identified and measured in research studies they can then be incorporated in predictions for practical noise control.

PREDICTION OF NOISE NUISANCE

We have now covered, fairly comprehensively, the large number of problems which arise from trying to relate nuisance to traffic noise, ranging from scale construction and noise units to sources of error and intervening variables. Now, although this is an essential requirement if we are to have a general understanding of the problem of trying to predict noise nuisance, it does not in itself take us as far as actual predictions.

In looking at the matter from the standpoint of a social and community problem which we want to ameliorate, even if we cannot entirely solve, we can see three distinct stages, or sub-problems. One is the attempt to relate measured nuisance to its presumed cause; namely, traffic noise. The second is to transform this demonstration of relatedness to quantitative predictions. The third is to formulate from these, standards of acceptable noise climate. This final stage extends beyond the present chapter, since it involves a political decision-making process and requires estimates of economic and social cost. But before such a process can occur on other than a purely intuitive basis, we have to deal with the second stage—that of predicting noise nuisance.

The procedures so far discussed only tell us whether or not nuisance is significantly related to traffic noise. They do not tell us precisely in what way it is related; that is to say, how much noise produces how much nuisance. A straightforward statistical test of associativity (such as the X^2 test) can tell us that X is associated with Y. A test such as product–moment or rank-order correlation can tell us that a change of X is associated with a change of Y. The extent to which these covariances are associated is measured by the coefficient of correlation, from -1 to 0 (inverse) or 0 to 1 (positive), which is therefore a statement of probability. At no point does it state quantitatively how much change of X (i.e. noise nuisance level) occurs with the observed change of Y (noise level).

To do this we must pass from correlational to regression analysis and this is a hazardous step to take. Since correlation is a process of associating dynamic values, covariances, it is concerned essentially with processes of change. But it is not essential that these changes be viewed as metric, as corresponding to a linear or other functional scale. They could be treated as merely a series of states, one succeeding the other. In this case we could treat them by rank-order correlations, which do not require any assumptions about the metric of items but only their serial order or ranking. Only if we can assume that the items are metrically related are we strictly entitled to use product–moment correlation. When we want to measure the change in one parameter relative to the associated amount of change in another, we can no longer make this choice, since such analysis itself presupposes a metric. There can be no meaning to measurement as a procedure without a scale of measurement. But without taking this step we cannot pass from relating nuisance and traffic noise to predicting how much nuisance traffic noise produces.

This problem has received much attention from statisticians, and many social studies of topics other than traffic noise have attracted censorious comments and critical arrows from experts on statistical analysis and design. Here we can only recognise that the problem exists and indicate the sort of answer that has been put forward. In strict terms it has to be admitted that only a non-parametric form of regression analysis—a contradiction in terms—would altogether eliminate it.

To make practical predictions, the gap between ordinal and cardinal numbers has to be jumped. And from sheer necessity this requirement overrides that of certainty for all the assumptions on which our procedures rest. We must therefore begin by postulating that the scales on which dissatisfaction or annoyance are recorded possess more than ordinal significance—that they have extension. If we accept the classical demonstrations of Thurstone[38], as regards the invariance of kth intervals in a numerical or semantic scale, and the validity of Guttman test procedures for the consistency and distancing of such items, we can go on to postulate that a scale which meets the requirements of these tests will not depart markedly from an interval scale. It is in order to point out that these assumptions have been made for the past 50 years, in correlating intelligence tests and examination results, without any disastrous results. A further justification for applying such procedures to annoyance scales is that when annoyance scores are examined in a number of surveys the distributions for the total population are seen to be approximately normal. As the scale on one dimension, that of measured noise, is already defined as an interval scale it raises no problem. In practice, therefore, all noise nuisance surveys have, after applying test procedures to the psychological scales, proceeded to use product–moment correlation and regression analysis.

It does not, of course, follow that a linear regression is implied, and a case in point here is Keighley's study[13] of office noise, which did not rely on linear regression but applied a Probit transformation to the dissatisfaction scores. In practice, however, the use of linear regression is common merely because it represents the simplest assumption about the form of the relationship, an assumption which can always be extended if necessary; that is, if the data cannot be made to fit such a regression. Such a case is discussed by Robinson,[27] considering the probably sigmoidal form of the noise-annoyance curve for aircraft noise. Even if the form of the regression equation is not linear, it is possible that over the actual range of conditions encountered, in the case of traffic noise rarely more than about 20–24 dBA, the departure from linearity is insufficient to result in significant errors.

It has therefore been assumed that a linear regression equation can be set up of the form $L = A + B + N \ldots$, or that a regression line may be generated by calculating the line of best fit to the plotted points for noise level and annoyance by the method of least squares. The limits of confidence

TABLE 2.2

RESULTS OF THREE SURVEYS CORRELATING NOISE NUISANCE WITH NOISE LEVEL

Feature	Paris, 1971	London, 1968	Stockholm, 1968
Sample	693	1120	326
Noise range, dBA L_{50}	53–75	53·5–68	43–67
Best results, units	$L_{50}\ r = 0·84\ (0·32)*$ $L_{10}\ r = 0·84\ (0·32)*$ Leq $r = 0·84\ (0·32)*$	TNI $r = 0·88\ (0·29)*$ 18 h $L_{10} = 0·8$ —	Leq $r = 0·96$ $L_{10}\ r = 0·88$ $L_{50}\ r = 0·82$
Noise level at 50th %ile in L_{50}	69	61	56 (Disturbed only)
Suggested limit	Below 70 dBA L_{50}	68 dBA 18 h L_{10} or 74 TNI	59 dBA L_{50}
Type of scale	7 pt. Dissatisfaction + interference with activities	7 pt. Dissatisfaction	10 pt. Category index of bother and exposure
Type of traffic	Mainly free flow, some congestion	Free flow, no congestion	Free flow, little congestion

* Individual correlation.

about such a line may also be calculated and are usually fixed so that 95% of the observations fall within these. Alternatively, the precision of the prediction may be indicated by the interquartile limits, the boundaries within which 75% of scores are located, or by limits at one or more standard deviation from the line.

Table 2.2 gives some of the results of the three surveys which have been most successful in correlating noise nuisance with noise level, although comparison of their predictions is not easy since each study has employed a different noise measure and the nuisance scales do not in all cases correspond.

Nevertheless, some comparison is possible, although it can be neither precise nor very reliable. Thus it has been necessary to convert the noise units employed to obtain the best results in each study to a single value, L_{50}, and this is a somewhat hazardous procedure. Again, the results of the French study are those without 'correction' for neighbourhood satisfaction, although they do contain a term for physical exposure while the British results do not. The Swedish data are based on noise values corrected for

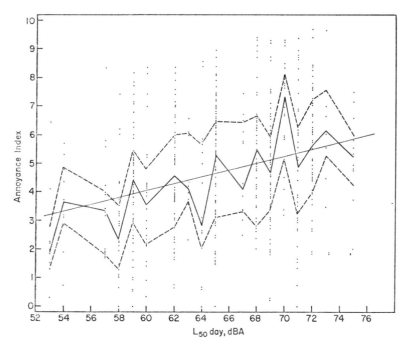

FIG. 2.6.   Annoyance related to median noise level (L_{50}).

distance of the dwelling, while the French and British results are based on
noise levels 1 m from the building facade. Finally, it is noticeable that the
range of noise levels covered by each study is somewhat different. Consequently, the 50th percentile, taken as the midpoint of the annoyance
scale, will be lower when the sample contains more sites at low noise levels.
Since it is not possible to make a completely satisfactory comparison of the
three studies, the best plotted results for each one are given in Figs. 2.6–2.11.

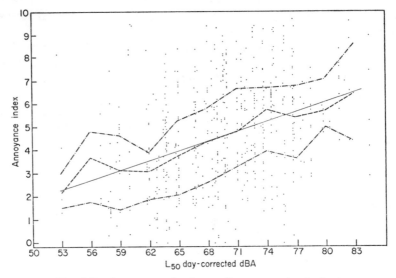

FIG. 2.7.   Annoyance related to median corrected noise level.

The plots for the Paris study (Figs. 2.6 and 2.7) show the effect of
'correcting' the data by means of the intervening variable 'satisfaction with
the neighbourhood'. It will be seen that the deviation of mean scores for
each condition is reduced markedly, although there is little if any effect on
the interquartile range, indicating that there remains considerable variance
not accounted for by satisfaction with neighbourhood and most probably
due to differences in noise sensitivity and other unexplored variables.

In the case of the British study (Figs. 2.8 and 2.9) the two diagrams relate
the best results, one for 24 hours and the other for 18 hours. This is because
the 24-hour L_{10} failed to give a predictive result equally as good as the
Traffic Noise Index, while the 18-hour value did so. It is also noteworthy
that the 95% confidence limit is shown. It can be seen that the 'pilot' sites
fall outside these. Finally, the authors advocate the use of the lower limit of

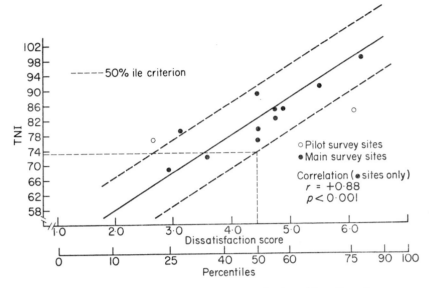

FIG. 2.8.   Relationship between TNI and dissatisfaction with traffic noise.

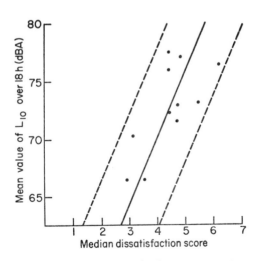

FIG. 2.9.   Regression of dissatisfaction on L_{10} (18 h value).

confidence in discussing standards, on the ground that if errors occur there is then only a 1 in 40 chance that these would have an adverse effect.

The Swedish plots (Figs. 2.10 and 2.11), like the British but unlike the French, show the position of actual site scores about the line of best fit, but show no confidence limits, although it is evident that the data would fall well within these. A curious feature is that the correlation coefficients

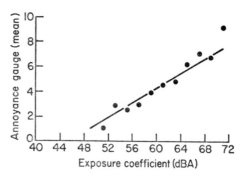

FIG. 2.10. Mean value of the measure of disturbance within 2 dBA classes with exposure coefficient E1 (mean energy value).

accompanying the plots give a lower value (0·87 and 0·91 as against 0·96) for the data corrected for distance, height and partial shielding than that obtained from the crude values. Although the method by which the regression formula is arrived at appears to be a straightforward equation of the type discussed above, the formula itself is not given, so that the weight of the various coefficients, or what these are, remains unknown.

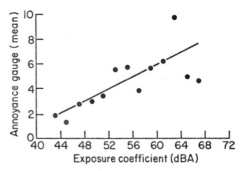

FIG. 2.11. Mean value of the annoyance gauge within 2 dBA classes with exposure coefficient E6 (50 % level).

Lastly, the range of annoyance occasioned by the actual noise levels may be compared. The Swedish conditions were on the whole quietest but the range of noise levels was 24 dBA, associated with a range of 60% of the annoyance scale. The British sites were slightly noisier but covered a smaller range. The French sites ranged in noisiness almost as widely as the Swedish, though at a much higher overall level. Nevertheless, the annoyance scores traverse only 40% of their possible range. On this basis it would seem that the Parisian respondents are least sensitive to noise and the Stockholm residents most, with London residents not much less so.

Despite differences between community standards and life-styles in the three capitals, there seems to be a fair degree of agreement between these results, if they are compared by looking at the noise levels at which 50% of the population are annoyed or dissatisfied, a level which has frequently been regarded as a guide to planning and noise control. Where the studies differ is with regard to what happens above or below this level, the extremities of the annoyance scale being reached more rapidly in some cases than in others.

There is also some difficulty in making very precise comparisons since the results are often expressed in different noise units not easily converted into common terms, or they relate to different limited periods of the day, or have been normalised for standard volumes of traffic such as 2000 vehicles per hour. In fact it would seem increasingly necessary that the results of noise nuisance surveys be expressed in a manner which facilitates comparison of acceptable noise levels. While such a requirement should not be allowed to impede the free development of the best prediction units, at least the results might be presented where possible in more than one unit.

With some loss in accuracy from translation of scales and noise units it is possible to make rough comparisons of these results. Using the median level (L_{50}) in dBA and the midpoint of the annoyance scale, roughly corresponding to the point above which half the population are disturbed, this level is reached, in the case of motorway noise, at about 64–65 dBA for Paris[23] and at 62 dBA for London.[14] Both these studies suggest a sharper perception of annoyance than the later results for Paris city traffic[15] which yields a figure of 68–69 dBA. The Swedish study yields a more critical result, taking the average of those disturbed and 'seriously' disturbed; the 50% level is reached at around 67 dBA Leq, which corresponds roughly with the results obtained in the London study.

The critical level for traffic noise would therefore appear to be around 65–70 dBA (L_{50}), a figure above which it is certain that a majority of the population will be seriously disturbed—a conclusion arrived at by

Aubree.[15] The results of the London survey, however, suggest that this level is excessive if it is intended to base standards on survey data. Thus the Urban Motorways Committee appointed by the British Government[39] suggested that a level of 70 dBA L_{10} (roughly 63–64 dBA L_{50}) during daytime be regarded as an absolute upper limit and one much above what was desirable. There is then a difference, perhaps exceeding 5 dBA, between the levels which different studies suggest as acceptable, yielding values ranging from 60 to 65 dBA average level.

Although the degree to which traffic noise interferes with various activities has been used to calibrate annoyance scales, and the studies show that they correlate highly with annoyance, relatively little information has been published on the direct relation between these effects and noise levels. The only figures which can be picked out distinctly are those which indicate that, at about 64–65 dBA, half the residents interviewed say that they would like to move if given the opportunity. At the same level, half claim that they have to close windows in warm weather when entertaining guests, although this rises to about 69 dBA for occasions when the family are alone. Finally, all studies appear to agree in classifying the relative disturbance occasioned by different noise sources, the worst being motor cycles, followed by heavy vehicles, sports cars and, least bothersome of all, ordinary private cars.

It is evident from what has been said, as regards the factors which need to be taken into account in predicting nuisance, that regression equations will tend increasingly to become standardised, with similar terms being introduced to account for the various physical and psychological elements. This being so, it is worth while at this point to list the various psychological, social and physical factors and consider what part they can play in predicting nuisance.

Human Response Variables

1. The basic scale of dissatisfaction or annoyance, whether this is a composite index, a Guttman-type verbal scale, a semantic differential or even a simple three category Yes/No/Neither question.

2. Those aspects of activity which are interfered with by noise measured by questions which intercorrelate with the scale of dissatisfaction and calibrate it. These will include things like reading, conversation, watching TV, entertaining guests, opening windows in warm weather or when retiring at night, and so on.

3. The degree of noise sensitivity shown by different individuals, measured by some test whose score can be entered as a term in the equation.

4. The degree of exposure depending upon the pattern and style of living, whether a person is at home much or little and what he or she does at home.

5. The degree of satisfaction with the local environment, whether obtained by a direct scale of appraisal or from answers to particular questions.

It is, of course, conceivable that other features could be incorporated, such as the attitude to traffic or environmental noise itself. But it is hardly desirable to extend the list overmuch, partly because with decreasing residual variance the gains are increasingly marginal and partly because, as the complement to this, increasing the number of terms in the multiple correlation or regression equation raises the level of significance required for effective prediction.

Physical Variables

1. The noise level, measured in various ways such as the peak, average or background levels, or any combination of these. Or, alternatively, some unit such as Leq or LNP which combines the mean value with the fluctuating components.

2. Indirect measures of noise level, correlating with it, such as the volume of traffic or the number or proportion of heavy vehicles, or some combination of these parameters.

3. The physical exposure factor related to the construction and design of the building, or the location of the principal room and the possibility of making this one shielded from road noise.

Now when all these elements, both social and physical, are introduced into an equation, whether by a stepwise process of canonical correlation, or by a simultaneous multiple regression, the result is to produce a 'best possible' prediction with each item contributing to the total variance according to the degree of importance it possesses in the whole situation. Of course, what is being described here is the outline of a computer operation and it should be evident that it is only the advent of powerful computers with large storage and capability for complex programming which has made possible types of analysis which were formerly of interest only to students of advanced statistics.

It is also necessary to contemplate the possibility that the items which are included on the physical side of the equation may not be limited to purely acoustic units or to correlative measures such as traffic flow. We cannot assume that a single noise measure, even one which combines different items of acoustic information such as TNI, Leq or LNP, will adequately represent all the variations which arise from the variety of flows and patterns of urban road traffic.

Work by Langdon suggests that optimal regression equations representing differences of this kind, while remaining invariant in form, will tend to vary with respect to the 'weightings' allotted to the various physical constituent items. Thus where traffic is free-flowing the noise term tends to play a dominant part in prediction. Where it is disordered and congested, the volume of traffic and the proportion of heavy vehicles become preponderant factors. Such a possibility appears to have been anticipated by Aubree,[15] who concluded by conceding that annoyance generated by ordinary city traffic may be different from that occasioned by motorways, being more complex and the resultant of a greater number of variables. Nor is there anything particularly novel in such a combination of physical variables, for it is anticipated in the Noise and Number Index developed in 1963 which has terms for peak noise and vehicle flow.

Such is the type of 'model' which seems to be required to produce an accurate quantitative prediction of nuisance caused by traffic noise. Yet much of this required information is available only through a social survey. Hence prediction of nuisance for the purpose of setting community standards may be based on all the factors mentioned above. But in deciding whether or not nuisance is likely to arise at a given spot and, if so, how much, the prediction will be one limited by the noise, traffic flow and exposure data available, as related to the basic noise nuisance scores. All other data, such as satisfaction with the locality, the pattern of living and the degree of noise sensitivity, are excluded from this prediction, since they cannot be known at any particular site where the population will be one varying in habits, sensitivity and environmental satisfaction.

It is therefore essential to distinguish between 'predictors' developed within a research study and 'predictors' available for practical application. The former are absolutely essential if a 'model' is to be developed which will account for as much of the variance as possible. The predictors used when the research results are applied in practice will be fewer and the predictions correspondingly less accurate, but they will have been derived from this conceptual framework. So we shall have some idea of what is happening rather than judging the situation analogously, like guessing the shape of an iceberg from the part we can see.

As a generalisation we can say that improvements to the physical measures will tend to raise the correlation coefficients between these and *group* scores; that is, the average amount of nuisance at particular sites.

This is because the effect of different physical measures is to alter the ordering of the various sites along the noise continuum. It is the effect which can be seen in the illustrations comparing nuisance measured by L_{10} and TNI (Figs. 2.8 and 2.9).

On the other hand, correlation of noise with *individual* scores will be little affected by such a procedure, since the major portion of the residual variance is due to the dispersion of individual scores at any given site. Individual correlation coefficients will only be raised by improvement in scaling of human response, mainly by information which takes account of the individual differences responsible for the dispersion of annoyance scores.

The effect of substituting different types of noise measure has been to increase the variance accounted for by a marginal degree, and if expressed in terms of individual correlation this appears minute. The improvement derived from more detailed social information is on the other hand considerable, resulting in significant increases in correlation values.

From this we should not conclude, however, that it is in this direction that improvement in prediction should be sought. For in the first place, most of the social information cannot be used in practical prediction since it cannot be known in the absence of a social survey. In the second place, it is of little practical use even if it were known, for any given site will contain a mixed population. Unless we are prepared to advocate that populations should be selected, sending those who are not sensitive to noise to busy localities, sensitive people to quiet spots, the prediction cannot apply very accurately. And this applies equally to all the other individual differences between people and families.

As we shall be trying to establish standards for whole communities and trying to apply these to groups of people living together at different places it would seem that the best guide to selecting predictors ought then to be the correlation between noise level and the median site score. So long as this is based upon a well-constructed model, in which we have been able to account for the greater part of total variance, we can continue to accept the quite high prediction accuracy for groups and overlook the extent to which individuals diverge from the group score. On this basis it would seem that all the recent surveys of traffic noise which have been described are capable of yielding nuisance measures with real predictive value. And it would appear that small improvements on these already respectably high correlations can be achieved by the introduction of maximum information about the noise level and the traffic pattern.

To speak of 'overlooking the extent to which individuals diverge' from the group score would at first sight appear to indicate a rather cavalier attitude to human suffering, and in a series of recent papers Bryan et al.[40,41] have suggested that the consequence of predicting nuisance from group scores and basing community standards upon them is to expose those more sensitive to noise to unacceptable levels.

Now if the best fitting line to group scores is used as the basis for community standards then the danger to which Bryan and his colleagues have drawn attention is a real one, especially given the wide dispersions typical of noise nuisance surveys. But while claiming that the needs of vulnerable sub-groups of the population require special consideration, Bryan fails to make concrete proposals as to how this might be given. It would seem, however, that the simplest answer to the problem is that which has already been suggested by Langdon and Scholes,[42] namely, to base nuisance standards and consequent predictions on the lower limit of confidence (see Fig. 2.9). As it is the respondents sensitive to noise who are themselves responsible for the dispersion scores on this side of the best fitting line, their needs will automatically be taken care of by relating any standard to this.

NUISANCE MEASURES AND NUISANCE STANDARDS

Having regard to the derivation of noise nuisance measures from the correlation of group scores with noise levels, on the grounds that it is to groups rather than individuals that standards and predictions will be applied, and because greater accuracy would affect individuals but not community groups, the question arises: how are nuisance standards to be derived from such measures?

Now first of all such a question implies that, by some process, standards could in fact be derived. But this will only be the case if the regression of noise with nuisance is significant. Even if the correlation between noise and nuisance is high, the variance of nuisance with respect to noise may be insufficient to yield the basis for a standard. This may be seen by reference to Fig. 2.6, taken from the Paris traffic noise study.[15] Although the correlation between mean scores and noise level is very high, the slope of the regression line with change of noise level is so feeble that a wide range of noise levels could be traversed within the 95% confidence limits. This means that there would be no certainty that any increase in satisfaction could be predicted for that reduction in noise level. In this case it is difficult to see how a community standard, even if it were established on the crude basis that below 60 dBA (L_{50}) a majority appear satisfied and above 70 dBA a majority appear dissatisfied, could be applied so as to produce a predictable increase in satisfaction at any given location. All that can be said is that the community as a whole expresses a degree of dissatisfaction with traffic noise which is significantly related to the noise level. But the rate at which this changes with noise level is insufficient to indicate the benefits of remedial measures unless they were undertaken on a drastic scale.

The communities covered by the Stockholm and the London studies appear to satisfy this condition. The slope of the noise–nuisance regression is steep enough so that a reduction of noise at any point along the scales of about 7 dBA would shift the annoyance score a greater distance than the width of the confidence limit. In other words, here it could be predicted that a noise reduction of even 5 dBA would produce a measurable improvement. To do this for Paris would, from the evidence, require a reduction of more than 10 dBA, something much more difficult to achieve.

Assuming that a standard can be derived from survey results, and here it is emphasised that this depends on the community responsiveness rather than on the quality of the research work, it is also necessary that such a standard is a practical one, an aspect that has already been mentioned. Here it is only necessary to stress that any standard must be couched in noise or traffic units which can be measured relatively simply and are amenable to calculation and prediction. This is a difficulty which arises with units such as TNI or LNP.

In the first place, the measurement of either requires the use of complicated and expensive apparatus. Secondly, there are very few tables or graphical plots which show how these units attenuate over distance, how they are affected by acoustic barriers or the walls and windows of buildings (see Chapter 5). These objections are even more serious in the case of LNP, since the major reason advanced for its use is that it may be applied to any source of noise whether road or rail traffic or aircraft noise, or any combination of them occurring simultaneously, in which case it would be necessary to be able to predict the transmission and attenuation of noise from each or all of these sources in the common unit, something which acousticians are very far from being able to do at present.

These considerations, then, may be said to define the necessary conditions on which a noise measure intended for community standards would need to be based. It would also be desirable to be able to distinguish between levels regarded as acceptable for daytime and for night, and for small towns as distinct from busy urban areas. Assuming that a unit meeting these requirements, such as Leq over 24 hours, or L_{10} over 18 hours, supplemented if necessary with traffic data, were to be used in a situation where community attitudes are responsive enough to guarantee benefits from the scale of noise reduction which could be achieved, how is such a standard to be derived?

It may be as well to realise immediately that a standard cannot be derived directly from scientific data. All that this can show is a relationship between traffic noise level and the degree of annoyance, or if this is transformed into a percentile scale, the proportion of the population annoyed at a given noise level. It cannot say what is a desirable level. It has sometimes been suggested that the 'reasonable' level is that intercepted at the 50th percentile, either for the regression line or the lower limit of confidence, on the grounds that this represents the level which the 'average' person will find acceptable. But there seems no basis for this other than a vague gesture to Benthamite Utilitarianism.

To put forward any noise nuisance standard involves making a conscious decision about the permissible level of noise. The most that the scientific evidence can do is to demonstrate as clearly as possible the consequence of this in terms of human attitudes and social behaviour. And this is not made easier by the fact that none of the traffic noise surveys has so far indicated precisely at what levels of noise the various daily activities are interfered with and to what degree.

The choice of a community standard for controlling traffic noise is, in practice, limited by two main considerations: that the amount of nuisance should not exceed a level acceptable to some proportion of the population, and that the cost of achieving such a standard should not be too great, either directly or by undue reduction of the efficiency of the transport system. Such a standard is therefore a compromise between what is desired and what can be afforded. It is therefore a question for public discussion and the deliberation of administrative bodies, and as such falls outside the scope of social science. But if the social scientist is to provide useful material for such discussion there remain a number of problems to be solved.

Our noise nuisance measures must be rendered more precise. Attitude scales must be improved, which means we must know more about what they are measuring, how scores are affected by individual differences and by how much. And the scales must be comprehensively calibrated. We have to be able to define annoyance and dissatisfaction in concrete terms, to be able to say just what traffic noise means in terms of difficulty in reading, listening, talking, entertaining friends, making music and all the other activities that make life worth living, and then trying to get to sleep at the end of the day.

The psychological and behavioural measures we have evolved need to be linked with, and so far as possible, cross-checked with economic data. This may take the form of direct monetary evaluation, possibly related to other environmental nuisances and amenities which can be compared with consumer evaluations of normally purchased goods and services, or it may develop through studies of the economic consequences, not merely prices and rents, but direct and indirect expenditures related to traffic. We must also study traffic noise over a more varied range of conditions. We need to assess its effects and develop effective units applicable not only to motorways and relatively simple orderly flows, but the range of urban conditions including those with high and varying proportions of heavy vehicles, congested and disordered flows, light-controlled intersections and pedestrian movements, all of which disrupt any simple noise-flow relationship.

Noise from road traffic must be related to that from other sources, such as railways, aircraft and industry, whether through the use of a common unit or by developing methods for comparing specific measures. Finally, there is a dearth of information on the specific effects of traffic noise on people in buildings other than dwellings, in particular in schools and hospitals.

Although it is clear from the above that there remains a large programme of research which must be undertaken if we are to have anything like a comprehensive understanding of the effects of traffic noise, and arising from this a set of reliable standards, it should also be evident from what has already been done that sufficient knowledge exists to enable a beginning to be made on planning and administrative measures.

Such measures, which are necessarily of a temporary and compromise nature, are exemplified by the British Land Compensation Act of 1973, which imposes a limit of 68 dBA L_{10} to external noise on all new road developments, provides compensation for loss of amenity or alternatively reimburses the resident the cost of acoustic treatment. These measures, which will be discussed later in more detail, represent a response to public demand and are a direct outcome of scientific information provided to expert bodies such as the British Noise Advisory Council, through which public policy adopted by government is largely decided.

Thus, while a comprehensive review shows the difficulties of the problem and how much yet remains to be solved, it also demonstrates that there is already ample information on which to base immediate, if interim, standards of noise control.

REFERENCES

1. Fog, H. and Jonsson, E. 'Traffic Noise in Residential Areas', Report 36E, National Building Research Institute, Stockholm, 1968.
2. 'The Social Impact of Noise', U.S. Environmental Protection Agency, Hearings in Atlanta, 1971.
3. Wilson Committee Report, 'Noise', Command Paper 2056, App. XV, HMSO, London, 1963.
4. Borsky, P. N. In Transportation Noises, J. D. Chalupnik (Ed.), University of Washington, Seattle, 1970.
5. Parkin, P. H., Purkis, H. J., Stephenson, R. J. and Schlaffenberg, B. The London Noise Survey, HMSO, London, 1968.
6. McKennell, A. C. In Transportation Noises, J. D. Chalupnik (Ed.), University of Washington, Seattle, 1970.
7. Borsky, P. N. 'Community Reactions to Air Force Noise', Pts I and II, WADD Technical Report 60-689, AF 33 and 41, 1961.
8. McKennell, A. C. 'Aircraft Noise Annoyance around Heathrow Airport', S.S. 337, HMSO, London, 1963.
9. Guttman, L. 'A basis for scaling qualitative data', Amer. Social Rev., 9, 139–50, 1944.
10. Edwards, A. L. and Kilpatrick, F. P. 'Scale analysis and the measurement of social attitudes', Psychometrika, 13, 1948.
11. 'Community Reaction to Airport Noise', Final Report, NASA Contract NASW 1549, Tracor Corpn., Austin, Texas, 1970.
12. The Second Survey of Aircraft Noise Annoyance around London Airport, Dept. of Trade & Industry, HMSO, London, 1971.
13. Keighley, E. C. 'Acceptability criteria for noise in large offices', J. Sound & Vib., 11 (1), 83–93, 1970.
14. Griffiths, I. D. and Langdon, F. J. 'Subjective response to road traffic noise', J. Sound & Vib., 8 (1), 16–32, 1968.
15. Aubree, D. Ètude de la Gêne du au Trafic Automobile Urbain, CSTB, Paris, 1971.
16. Osgood, C. E., Suci, G. J. and Tannenberg, P. H. The Measurement of Meaning, Univ. of Illinois, Chicago and London, 1957.
17. Guilford, J. P. Psychometric Methods, 2nd Edn, McGraw-Hill, London, 1954.
18. Morgan, J. N. and Sonquist, J. A. 'The Determining of Interaction Effects', Technical Monograph No. 35, Survey Res. Centre, Univ. Michigan, 1964.
19. Botsford, J. H. 'A simple method for identifying acceptable noise exposures', J. Acoust. Soc. Am., 42, 810–19, 1967.
20. Botsford, J. H. 'The Weighting Game', Presented to 75th meeting Acoust. Soc. Am., May 1968; J. Acoust. Soc. Am., 44 (1), 381, 1968.
21. Alexandre, A. 'Prévision de la gêne due au bruit autour des aeroports et perspectives sur les moyen d'y remedier', Doctoral Thesis, University of Paris, April, 1970.
22. McKennell, A. C. and Hunt, E. A. 'Noise Annoyance in Central London', S.S. 332, HMSO, London, 1966.

23. Lamure, C. and Bacelon, M. 'La Gêne due au Bruit de la Circulation Automobile', *Cahiers du CSTB*, No. 88, Cahier 762, 1967.
24. Bruckmayer, F. and Lang, J. 'Disturbance of the population due to traffic noise', *Öst. Ing. Zeitschr.*, No. 8, 302–6; No. 9, 338–44; No. 10, 376–85, 1967 (in German).
25. Aubree, D., Auzou, S. and Rapin, J. M. *Ètude de la Gêne du au Trafic Automobile Urbain*, CSTB, Paris, 1971.
26. Griffiths, I. D. 'A note on the traffic noise index and the equivalent sound level', *J. Sound & Vib.*, **8** (2), 298–300, 1968.
27. Robinson, D. W. 'The Concept of Noise Pollution Level', Aero Report Ac 38, National Physical Laboratory, England, 1969.
28. Lamure, C. and Auzou, S. 'Les niveaux de bruit au voisinage des autoroutes dégagées', *Cahiers du CSTB*, No. 71, Cahier 599, Dec. 1964, Paris.
29. Christie, A. W., Prudhoe, J. and Cundill, M. A. 'Urban Freight Distribution: a Study of Operations in High Street, Putney', Transport & Road Research Laboratory, Report LR 556, 1973.
30. Aubree, D. *Ètude de la Gêne du au Bruit de Train*, CSTB, Paris, June 1973.
31. Troy, P. N. *The Quality of the Residential Environment*, Australian National University, Canberra, 1970.
32. Finney, D. J. *Probit Analysis*, Cambridge University Press, England, 1962.
33. Bryan, M. E. 'Noise laws don't protect the sensitive', *New Scientist*, **59** (865), 738–40, 1973.
34. Jonsson, E., Kajland, A., Pacagnella, B. and Sörensen, S. 'Annoyance reactions to traffic noise in Italy and Sweden', *Arch. Environ. Health*, **19**, 692–99, 1969.
35. Lawson, B. E. and Walters, D. 'The Effects of a New Motorway on an Established Residential Area', Presented to 3rd Conference Arch. Psychol., University of Surrey, September 1973.
36. Franken, P. A. and Jones, G. 'On response to community noise', *Applied Acoustics*, **2**, 241–46, 1969.
37. Cederlöf, R., Jonsson, E. and Kajland, A. 'Annoyance reactions to noise from motor vehicles: an experimental study', *Acoustica*, **13** (4), 1963.
38. Thurstone, L. L. 'Theory of attitude measurement', *Psychol. Rev.*, **36**, 222–41, 1929.
39. 'New Roads in Towns', Report of the Urban Motorways Committee, HMSO, London, July 1972.
40. Moreira, N. M. and Bryan, M. E. 'Noise annoyance susceptibility', *J. Sound & Vib.*, **21** (4), 449–62, 1972.
41. Bryan, M. E. and Tempest, W. 'Are our noise laws adequate?', *Applied Acoustics*, **6** (3), 219–32, 1973.
42. Langdon, F. J. and Scholes, W. E. 'The Traffic Noise Index, a method of controlling noise nuisance', *Arch. J.*, **147**, April 17, 1968; Building Res. Stn. Current Paper 38/68, London, 1968.

CHAPTER 3

THE SOCIAL COST OF NOISE
A. ALEXANDRE and J.-Ph. BARDE

The fact that noise has adverse effects on man and results in an overall reduction in his well-being means, in socio-economic terms, that noise has a *social cost*. It would seem natural and logical that this social cost be compensated for, and indeed the theory of welfare economics tells us that any economic agent responsible for any kind of cost should, at the same time, be responsible for the compensation of those costs it imposes upon society.

Every economic activity has a direct monetary counterpart stemming from some voluntary exchange. For example, the purchase of a pair of shoes is a voluntary act and such an exchange is possible through the medium of income earned by work, i.e. the sum of money (wage) paid to compensate the worker's effort. All such transactions are carried out in a market, i.e. a meeting place for buyers and sellers of products (goods and services), production factors (capital, labour, land).

Furthermore, as any exchange is assumed to be voluntary and motivated by each person's desire to maximise his own self interest, classical economic theory concludes that all interests are in harmony. For example, the car manufacturer produces goods useful—or at least considered as such—to the community, with the result that his own private interest and that of the community are in harmony. In carrying out his activities, moreover, he pays compensation for the production factors used—wages for labour, interest for capital, etc.—while his production costs include the other inputs such as the raw materials used. The price of these factors reflects their scarcity (if infinite quantities were available these factors would be free—except for extraction and transportation costs) and prevents their being wasted. Due to these mechanisms of compensation through the medium of prices, *the private cost and social cost of each activity are said*

70

to coincide. For example, the social cost of labour is compensated by a wage, which is a component of the employer's private cost; the social cost of using steel is compensated by its purchase price, and so on. Unfortunately, no such ideal harmonious state of affairs exists, since economic activity is accompanied by a growing number of *uncompensated social costs*. Building sites are a daily source of noise, subjecting both workers and neighbouring residents to a series of disamenities which logically should be included in the building contractor's prices. But this is rarely the case and the fact that a contractor may freely avail himself of the right to make noise, i.e. to affect the community adversely without paying compensation, in no way induces him to account for the social cost of his activities in the prices he charges for his buildings. Again, people living in the neighbourhood of building sites or traffic areas are unilaterally subjected to noise without any voluntary transaction taking place. In economic terms, therefore, these activities may be said to be associated with *negative flows* which occur *outside the market* (without any voluntary transaction). The negative flows will be called *external diseconomies* or *negative external effects* (an activity may also provide unquantified benefits which therefore may be termed *positive external effects*).

If this 'right' to make noise were given a *price* which had to be paid by the agent making the nuisance, market decisions would reflect the social cost of noise and the noise would be at a level commensurate with this price. In other words, if the price of noise resulted from market mechanisms we should be aware of the value society places on noise—and hence on silence—and the economic system would automatically '*internalise*' the cost of noise, so that the noise would be kept to an optimum level reflecting social preferences.

In Fig. 3.1, curve S represents the marginal social cost of noise, i.e. the increase in disamenity caused by each additional unit of noise. Curve R (marginal cost of noise reduction) represents the cost of reducing each additional unit of noise. The greater the noise, the higher the social cost(*S*) The lower the noise level, the higher the cost of reducing noise by a further unit. The characteristic shape of the curves will be noted: the shallowness of the curve S at the beginning reflects the fact that a low noise level causes little annoyance and hence the social cost is small. As the noise increases, however, the additional decibels become more annoying. Following curve R from *RIGHT* to *LEFT*, it can be seen that an initial reduction in the noise level is easy to achieve and therefore costs little whereas, conversely, it would be very costly to eliminate the noise entirely, thus illustrating the law of diminishing returns. It should also be noted that these arguments are valid only if the social cost can actually be measured in monetary terms.

Curves S and R can also be interpreted as supply and demand curves, R being the supply curve of silence and S the demand curve for silence. The 'price of silence' is thus fixed at level P* corresponding to a noise level N*.

There is clearly no advantage in reducing the noise beyond point Q, since the cost of reduction is then greater than the benefit to the community (benefit = reduction in social cost). If the noise is reduced to N, the corresponding social cost (S_1) would remain considerably ~~higher~~. *LOWER*

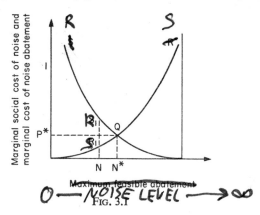

FIG. 3.1

For the economist, N* is then the optimum level of noise, since it is here that the social cost is compensated, without exceeding the abatement cost.

Now the problem is that noise is an external effect, i.e. it is external to the market, hence *we ignore the price of noise*, that is to say the value society is putting on noise (silence). Ignoring the curve S, we cannot determine the point N* which determines the noise level society is prepared to accept. If, for instance, public authorities were to decide to set a standard for automobile noise emission, which criteria should be used to set this standard? If it were set at a level lower than society would accept, the cost of noise abatement will be higher than the corresponding benefits. If the set level were higher than that which society would demand, the social cost would exceed the quantity society is prepared to accept.

It is then of paramount importance to assess the social cost of noise in order to have at our disposal some criterion for decision-making, particularly for the setting of noise standards.

Since the market does not provide any help, we must turn to other methods to calculate the social cost of noise.

Under the general heading of 'cost benefit analysis', several attempts have been made to calculate this social cost. The principle is to estimate the curves S and R, the costs and benefits of noise abatement. If we assume we know the cost side of the equation; that is, the cost of abating noise, the second, and the most difficult, part of the task is to assess the benefits of noise abatement. Benefits are defined as the damages avoided due to the noise reduction. Thus, the calculation of benefits implies the calculation of social costs. The problem is twofold; first one must determine the different costs and, secondly, one must try to measure those costs by putting a money value on them.

THE DETERMINATION OF SOCIAL COSTS

The phenomenon of 'externalities' as described above, can have a number of different dimensions, namely:

1. Externalities between producers—the noise on a building site may reduce the productivity of employees in nearby offices.
2. Producers' externalities on consumers, e.g. all kinds of disamenities due to production activities (noise of lorries).
3. Consumers' externalities on producers—traffic noise has negative effects on workers.
4. Externalities between consumers—general case of traffic noise.

Clearly all of these cases apply to traffic noise. But, in practice, it is difficult to grasp the precise scope and meaning of the social cost of noise.

First, we should list the *direct* effects of traffic noise, e.g. a loss in the productivity of workers, difficulty in communication, physiological effects, effects on sleep, etc.

Second, the *indirect* effects of noise must be calculated. These may be even more important, although they are closely interrelated with the direct effects and thus difficult to separate. This is the case, for instance, if loss of sleep were to result in a reduced working capacity and/or in negative changes in a behaviour pattern. Or, if the noise level were to make it difficult for a teacher to communicate with his students, the social cost of the deterioration of teaching might be enormous—this has been demonstrated in a number of cases for schools located near airports,[1] where a decline in the working capacity of pupils was discovered. But how can we evaluate the social cost to a child of, perhaps, being denied entrance to a university?

Generally speaking, and apart from some quite direct physiological effects, the principal effect of noise is annoyance, the degree of which may vary widely between individuals. Sometimes the annoyance is not even directly felt, despite the existence of its effects. The general concept of 'amenity loss' is replete with value judgements as well as with uncertainties, so that it may prove extremely difficult to determine precisely the social cost of noise.

But this is not the real job of the economist, whose task will mainly consist in making a quantitative evaluation of these social costs.

THE QUANTITATIVE EVALUATION OF THE SOCIAL COST OF NOISE

If one wants to compare the benefits of noise abatement with the corresponding costs, one must find a common unit of measurement. As the abatement costs are expressed in money terms, the problem is then to put a money value on the social costs. Since there is no 'noise market', we must try to find some indirect means of evaluation. For this, a number of methods have been proposed and they can be roughly classified in two main categories: the use of indirect market indicators, and the direct estimation of the demand for noise abatement.

In the framework of this short chapter, it is impossible to make a detailed analysis of each. In order to give the decision-maker an idea of the possible methods we shall briefly explain their principles with a short list of their respective advantages and disadvantages.

It is worth mentioning that these methods are not exclusively applicable to noise but can be used to evaluate other types of environmental degradation (e.g. air and water pollution).

It must also be noted that the two methods are not mutually exclusive, but can be used *in a complementary manner*: one method may be better for evaluating one part of the social cost, whereas the other may yield better results for other costs.

The Use of Indirect Market Indicators

In some cases the effects of noise may result in monetary expenses which can be assessed from the market place. The most obvious case would be when physiological effects of noise lead to medical expenses, be it for medical treatment of deafness, or medication against headaches, or even sleeping pills. Of course, medical expenses do not accurately reflect the social costs of a disease. Clearly, they omit the costs of suffering and losses of welfare. Also, in some cases the medical care may not succeed in restoring an individual to the state of well-being in which he found himself before being exposed to noise. Therefore, all we can say is that, in some straightforward cases, medical expenses reflect part of the social cost of some noise effects, i.e. the amount of money necessary to treat the effect so that the individual can be in the same state of well-being as before the occurrence of the nuisance.

Along the same lines, another market indicator relates to *'amenity improvement expenditures'*, the expenditures of individuals to abate noise, e.g. double glazing of dwellings. Such expenditures might indicate the money people are willing to pay to get rid of the noise nuisance. But this case may not be as simple as it appears *prima facie*. Lassiere and Bowers[2] note that 'the motivation for installing double glazing is seldom related to noise alone, it is often done to improve thermal insulation as well, and secondly, the houseowner may put it in partly because he wants to increase the capital value of his property, and partly to reduce his indoor noise level'. But if one can assess precisely the expenditures which are allocated to noise abatement, they may represent a rough estimate of the private valuation of the social cost of noise.

If noise affects the productivity of a person there are some means to calculate these losses, especially if absenteeism occurs. In this case, costs can be evaluated through calculating income or profit reductions.

The most commonly used and most sophisticated method is the so-called *'housing market approach'*. The basic principle is quite simple: it is assumed that the value of a house not only reflects its building cost but also all the environmental features surrounding the house, such as proximity of green space, schools and shops, ease of access, air pollution and noise, etc.

Thus noise can affect the value of a house and if one can isolate the price decrease due to noise, it could represent an approximation of the social cost.

A number of housing market studies have been made, though as yet without any real significance or satisfactory results; some have yielded contradictory indications. In many cases the housing market has been studied around airports or traffic areas.

In the USA a study of Portland (Oregon) house rents, employing a stepwise regression of apartment rental values on several variables, including noise, was not able to assess any significant effects of noise on rents. If noise has negative effects on rents it is possible that these effects may be outweighed by the positive effects of other factors highly associated with increased noise, such as the proximity of schools, recreational facilities, access to highways, etc.[3]

Another study, this time in Toledo (Ohio), was no more conclusive. Values of properties situated near an expressway did not vary from those protected from noise. The most puzzling result came from a complementary survey of real estate agents and owner-occupiers. While the first group estimated that noise could lead to a 20–30% depreciation in house value, 63% of the residents living in an area with peak noise levels of 80–85 dBA declared that, if they were to repurchase, they would never choose a house near an expressway.[3] This suggests that even if noise is considered to be an importance nuisance, it is not strong enough to influence the housing market.

In fact, with this kind of study, where regression techniques are used, we are faced with the problem of mis-specification; that is, several variables influence the price of houses and it is difficult to account for all of them. Moreover, one can have strong doubts about the real relationship between noise and house prices. A recent survey of 340 houses in the suburbs of Birmingham, undertaken by the Statistical Research Unit of Keele University, has even suggested that house values increased with noise! The selling prices of houses situated along a noisy road increased at a greater rate than those of similar houses in a quiet road nearby. Obviously, noise does not itself increase house values, but other variables related to road proximity positively influence house prices. Furthermore, a social survey in this area indicated that, contrary to the Toledo case, the inhabitants were satisfied with living near the main road.[4]

It is worth mentioning that a number of housing market studies have been made around airports, principally within the framework of the Roskill Study for the siting of the Third London Airport. Positive correlations between noise levels and depreciation in house value were found; the results of the different studies are shown in Fig. 3.2.[5] The price variations were established by consulting samples of house agents. As these agents did not all give the same answers, the outcome was a rather wide dispersion about the average depreciation quoted. Furthermore, serious doubts have been cast about the real ability of house agents to estimate the impact of noise on house prices; it is also likely that they mix noise with other components of the environment. But it is quite possible that as airport noise is particularly intrusive, its impact on house prices is greater than other environmental factors, although the economic development around an airport leads, in fact, to an increase in the price of land. Should we then conclude that noise has no effects on house prices? All we can do is to make a few comments.

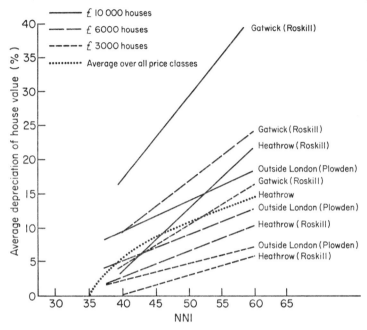

FIG. 3.2. House price depreciation percentages as a function of noise exposure.

First, the housing market is a very peculiar one, at least in some countries. A recent study in France has pointed out that so many constraints influence this market that it is not realistic to seek to relate noise to house prices.[6] Also, in many cases, one is confronted with a tight market, where demand greatly exceeds supply. A market with a marked scarcity of dwellings is less likely to reflect noise nuisance.

Second, the effects of this relative scarcity are far more pronounced for people with lower incomes. The Birmingham study applied to a homogeneous sample of low-income people. It must be noted that the higher the value of a house, the stronger the correlation with environmental variables. The price of an expensive villa will certainly be significantly affected by the construction of a motorway in front of it. But no effects are likely to be noted if the motorway is constructed next to apartment buildings occupied

by low-income people. And here we must point out an important consideration: the decision-maker may well decide to build a highway next to
low-income dwellings, since the social cost appears to be nil (or small)
compared to the social cost of the same operation in a wealthy residential
area. Such an example reveals a weakness of housing market studies which
are hardly likely to evaluate the real social cost of noise, if in reality noise
annoys the poor as well as the rich.

It is possible that in the future, with a non-scarcity housing situation
and when people have become more aware of and more sensitive to
environmental amenities, noise may have a significant impact on house
prices, because people will no longer be prepared to accept the noise
nuisance which they now endure.

Estimating the Demand for Noise Abatement

In order to estimate the social cost of noise, the Roskill Commission[7]
adopted a model in which four kinds of residents were defined:

	Social cost
1. Those moving because of airport noise	$S + R + D$
2. Those moving anyway	D
3. Those remaining	N
4. New entrants	zero

where, S = consumer's surplus defined as 'the difference between the
householder's subjective value of his house and the market value (. . .).
The subjective value (. . .) is the sum which the householder would consider
just sufficient to compensate him for loss of the property, assuming he had
to leave the noisy area altogether';[7] R = cost of moving; D = depreciation of house value; and N = the cost of enduring the noise.

Methods for evaluating D are those presented above. R is obtained from
the market. To evaluate S, questionnaires were used which asked the
householders what sum of money would compensate them for moving out
of the noisy area. N was estimated by relating annoyance scores to the
depreciation of house values (D).

In fact, the social survey technique consists in asking people the price
they would be prepared to pay (willingness to pay method) to get rid of the
noise, or the compensation they would require if they were to move
because of noise. In other words, one tries to make a direct evaluation of
the demand curve for noise abatement.

In a study by Plowden,[8] the author estimated N and S, which he designated as the 'endurance cost' and 'dislocation cost'.

The *endurance cost* 'borne by owner-occupiers who elect to stay in the area after the noise has been imposed', was estimated through a social survey. On the assumption that they were moving, people were asked to state the price they would be prepared to pay in order to buy the house they desired. Then, assuming different noise levels, people were required to say how much less the price of the house should be in order to compensate for the noisy environment. The results are given in Fig. 3.3 and indicate that a majority of people would not accept payment for the inconvenience of high noise levels.

The *dislocation cost* 'borne by each owner-occupier who decided to leave an area on account of the noise' was also estimated by survey techniques. Assuming they could sell their houses with a profit of £100,

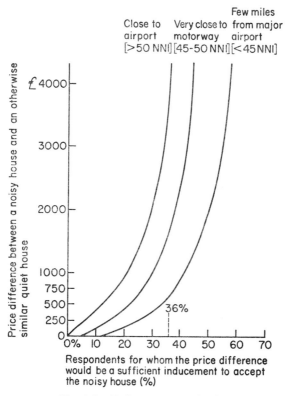

FIG. 3.3. Endurance costs of noise.

people were asked if they would then agree to move. If they did not agree they were asked what the difference in price would have to be to induce them to move. The results in Fig. 3.4 indicate that 38 % of the respondents *would not be prepared to move at any price.* Thus for these people moving would have an infinite social cost. Another survey conducted by the Roskill team found a proportion of 8 % 'infinites'. How should these 'infinite' values be treated? If they are to be considered at face value, a decision

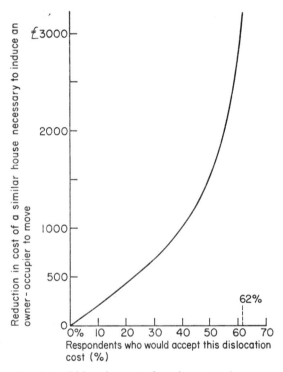

FIG. 3.4.   Dislocation cost of moving to another area.

should be taken not to build the airport. But in both Plowden's and the Roskill study, it was decided to assign an *arbitrary* value to these respondents. The METRA study set a cut-off figure of 50 % of the respondent's house value, while the Roskill team fixed an arbitrary value of £5000. Of course, these procedures cannot be justified. The existence of such infinite values may prove, on the contrary, that in some cases social cost cannot be compensated for by money.

Clearly the methods used appear to be questionable and indeed they have met with heavy criticism. To go into technical details would be beyond the scope of this chapter; rather we must focus on the fundamental criticisms of housing market studies and social surveys.

Main Criticisms of the Monetary Evaluations Methods of Noise Effects[9-11]

1. We have seen that the social cost of noise has many direct and indirect components, so that various *complementary* methods must be used in order to evaluate them. Thus the first question to be raised is: what is really reflected by a depreciation in house value (D) and to what extent? Assuming the existence of a correlation between noise levels and house prices, what is the precise meaning of this correlation? Does it reflect amenity losses, productivity losses, interference with sleep, effects on teaching in schools, etc.? To our knowledge, no precise definition has ever been given, and hence D seems to us to be an indicator of limited significance.

2. As already mentioned, a number of factors influence the housing market, and it is very difficult to isolate the effects of noise from the others. Furthermore, other factors usually have a much stronger influence than noise. Noise may be correlated with nearby public transport, access facilities, shopping centres, etc., which have a positive impact on house prices. Also, the construction of an airport or a motorway generally induces economic development of the area, causing a rapid increase in land values.

When people are asked whether they are annoyed by noise, they generally give a positive answer, especially if they are questioned about their preference for a quiet residence. But other parameters are decisive in the choice of residence: proximity of schools, place of work and shops; existence of a garage with the house, or of a certain type of central heating; bathroom, etc. It is therefore difficult to believe that people would move, leave their friends and change their way of life exclusively on account of noise. If noise were actually the decisive factor, areas around airports and highways would be deserted. People endure noise because they have no alternatives; this is probably the main reason why housing market models fail to evaluate the social cost of noise.

Last but not least, the hypothetical relationship between noise and house values may inhibit selling. If residents are told that noise is dangerous and affects their property prices, they may decide not to sell in order to avoid financial loss.

3. The influence of noise on property values reflects social losses all the less because people are often not aware of, or do not consider, the long-term effects of noise on health, life-style, etc. This fact biases answers to social surveys to a considerable extent, particularly since people are asked to answer questions quite rapidly.

4. Pearce has pointed out that 'the fundamental error lies in the un-recognised assymetry between social costs which have associated monetary flows and those which do not'.[9] On the one hand, people are referring to the *financial* values of their houses and moving costs; on the other hand, they are confronted with a noise nuisance which is not expressed in monetary terms. No relationship can really be established and people exposed to noise are not able to perceive any direct monetary benefit from noise abatement.

5. In social surveys, people are asked unrealistic questions. When they are questioned about the amount of money they would have to be paid in order to be compensated for a noise nuisance, they are free to throw out any figure they choose since they are put in a purely fictitious situation; and indeed this is what occurred in the METRA survey which showed that, for a number of people, moving would have *an infinite social cost*. This kind of approach is equivalent to asking people questions like 'What amount of money would you accept for leaving your friend?', or 'How much would be necessary to compensate you for changing the school and life-style of your children?'

6. As already mentioned, social factors play an important role: the rich man will give a high value to noise because he attributes a high marginal value to amenities and also because he is used to thinking in higher monetary terms than the poor man. Thus the rich will ask for large compensations which are not comparable with the nuisance values given by the poor. A delicate problem of weighting exists here.

7. So far, these surveys have completely neglected the losses to those people who live outside the noisy areas and who would prefer to live in the affected areas if only they were quiet. This 'option loss' could be important. But it is not at all certain that a monetary value could be placed on this loss. (This point has been suggested to us by D. W. Pearce.)

CAN WE EVALUATE THE SOCIAL COST OF NOISE?

At this stage it does not seem possible to give a monetary valuation of the social cost of noise. The techniques so far used have not been successful;

they yield contradictory results and leave room for value judgements on the part of those analysing the results.

Not only can the cost–benefit approach result in misleading conclusions; it may also be used to justify the hidden preferences of decision-makers. The variables employed are so vaguely defined that one could easily vary them to obtain a preconceived result or to justify a decision already taken. We are therefore of the opinion that in the absence of adequate methodologies, cost–benefit analyses, as applied to noise, constitute economically misleading, and politically dangerous, instruments.

We therefore need to use non-economic criteria for decision-making. The assessment of reliable annoyance scales has proved to be feasible (see Chapter 2) and a number of methods have been successfully used to reveal and rank individuals' preferences.[2,12]

The so-called '*priority evaluator*' is a gaming method, inviting people to allocate a sum of money between several alternatives for improvement of their environment under certain budget constraints. An electromechanical apparatus consisting of a board on which a set of variables is presented by small pictures is used. When an alternative is chosen the corresponding picture is illuminated.[2] Several experiments, in which the presentation of the problem and of the budget constraints has been varied, have yielded consistent results.

Simulation techniques have also been used. A sample of people is placed in a room where several environmental conditions are simulated (e.g. noise by loudspeakers). Respondents are asked to rate each of the simulated conditions.

To conclude, we can say that if the social cost of noise cannot be evaluated in monetary terms, it can be assessed through non-monetary indices such as annoyance scores (in the economist's jargon, these are called 'non-monetary damage functions', as opposed to 'monetary damage functions'). Also, public preferences can be ranked by means of gaming methods and simulation techniques. These certainly provide decision criteria. Of course, the decision-maker would not have at his disposal the purely economic criteria which determine the *optimal* noise level at which marginal social costs equal marginal abatement costs. But he would know fairly accurately people's preferences. From the annoyance scores, which relate noise levels to percentages of people annoyed, the decision-maker will know the exact number of people likely to suffer from noise (and to what extent they will suffer, if he takes a given decision). Non-monetary indices will also give an approximation of what are considered to be 'acceptable' or 'non-acceptable' noise levels, although they leave room for value judgements and uncertainties, due to the fact that people may not be sufficiently aware of the real noise impact on their well-being.

In any case, the choice that is ultimately made must be a political one. But if the non-economic approach loses in 'efficiency' by failing to assess the optimum, it certainly gains in justice and equity by stating people's preferences and levels of acceptability.

REFERENCES

1. US Environmental Protection Agency. *The Economic Impact of Noise*, Washington, 1971.
2. Lassiere, A. and Bowers, P. 'Studies on the Social Costs of Urban Road Transport (Noise and Pollution)', European Conference of Ministers of Transport, Report of the 18th Round Table on Transport Economics, Paris, 1972.
3. US Environmental Protection Agency. *The Economic Impact of Noise*, Washington, 1971.
4. Diffey, J. 'An Investigation into the Effect of High Traffic Noise on House Prices in a Homogeneous Submarket', presented at a seminar on house prices and the micro-economics of housing, London School of Economics, December 1971.
5. Opschoor, H. 'Damage functions, some theoretical and practical problems', in *Environmental Damage Costs*, OECD, Paris, 1974.
6. CRESAL. 'Les nuisances dues aux bruits de la circulation automobile urbaine: Effets sur le marché immobilier' (unpublished), Saint-Etienne, 1972.
7. Commission on the Third London Airport (Roskill Commission). Papers and proceedings, HMSO, London.
8. Plowden, S. *The Cost of Noise*. METRA, London, 1970.
9. Pearce, D. 'The economic evaluation of noise generating and noise abatement projects,' in *Problems of Environmental Economics*, OECD, Paris, 1972.
10. Paul, M. E. 'Can Aircraft Noise Nuisance be Measured in Money?', Oxford Economic Papers, November 1971.
11. Alexandre, A. and Barde, J.-Ph. *Le Temps du Bruit*, pp. 161–94, Flammarion, Paris, 1973.
12. Lagoueyte, J. P. 'Essai sur les Methodes d'Analyse et d'Évaluation des Preferences d'Environnement', Thesis, Université de Bordeaux I, France, October 1971.

CHAPTER 4

NOISE EMITTED BY ROAD TRAFFIC

C. LAMURE

MOTOR VEHICLE NOISE

Origins of Motor Vehicle Noise
The noise of a vehicle is caused by the engine, the tyres and the air turbulence. For urban traffic at low speeds engines are often the main sources of noise; tyres and road interaction contribute to noise at higher speeds and air turbulence is generally unimportant.

The noise due to the engine is produced by radiation of vibrating surfaces and by various individual sources such as exhaust, inlet manifold, transmission and cooling fan (Fig. 4.1).

Table 4.1 illustrates the principal components of the external and internal noise for a diesel truck.

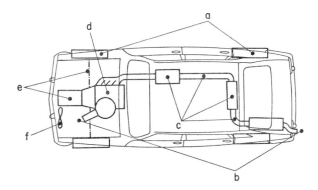

FIG. 4.1.   Major noise sources in a car. (a) Tyre noise; (b) primary inlet and exhaust noise; (c) noise radiated by inlet and exhaust systems walls; (d) engine vibration emitted noise; (e) gearbox and transmission noise; (f) cooling fan noise.

TABLE 4.1

COMPONENTS OF NOISE FOR A DIESEL TRUCK
(from Priede[1])

Origins of noise	Noise inside the vehicle	Noise outside the vehicle
Engine airborne noise and its transmission	Major source of low frequency noise	May be predominant, especially on trucks
Engine exhaust	Not important	Major source of low frequency noise
Engine inlet	Not important	Major source of low frequency noise following exhaust
Fan noise	May be noticeable	Can be significant in low and middle frequency ranges
Road-excited vibration	Major source of low frequency noise	Not significant
Road-excited tyre noise	Not significant	Significant

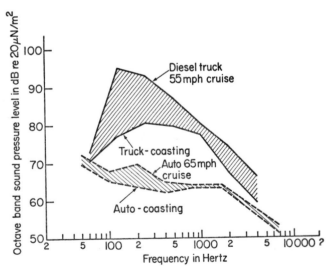

FIG. 4.2.   Diesel truck and automobile noise at highway speeds. Cruise and coasting
(at 15 m).

The spectra of the external noise at a distance of 15 m are shown for a ten-ton truck and for a car in Fig. 4.2. The noise emitted in the low frequency range is higher for trucks. The noise level decreases rapidly for frequencies above 2000 Hz. The corrected spectrum for A weighting frequency network shows that the most disturbing range is from 500 to 1000 Hz for cars and from 200 to 500 Hz for trucks.

Inlet and Exhaust Noise

When no silencing is used inlet and exhaust are predominant sources of noise. The exhaust noise level varies as the logarithm of the engine rotation speed. The increase of the noise level is around 45 dB per tenfold increase of engine speed.

A silencer reduces the induced pressure fluctuations and allows the air flow to communicate with the atmosphere through deflectors equipped with sound absorbent. With a silencer, the reduction of the exhaust noise level ranges from about 15 to 25 dB; the reduction for inlet noise is lower— approximately 10–15 dB.

Silencers or mufflers have a major effect on the noise produced, so that generally in trucks and often in cars the noise radiated by the engine masks that of the exhaust. For cars, very high quality silencers may provide a further reduction of only 5 dBA. Such silencers need space which is generally not available on cars, while specially designed resonators must be used to eliminate pure tones at low frequencies (under 140 Hz). The great insertion loss which occurs with a muffler of 36 kg is shown in Fig. 4.3(a). These types of large muffler generate back pressure and increase fuel consumption.

Turbocharging reduces noise by lowering the exhaust back pressure and also reduces the noise radiated by the engine. The attenuation of the high acoustical energy in the low frequencies is particularly important. Turbocharging is more and more used for engines over 200 hp. The reduction of the noise level due to turbocharging may be greater than 6 dBA, for the same output power.

Structure and Engine Noise

The noise radiated by the engine and the structure results from the vibration of flexible partitions (i.e. the sump). The level of noise emitted depends on the characteristics of existing forces and the size of the vibratory system in relation to the frequency of vibration. Following a number of investigations pursued by Priede at the Institute of Sound and Vibration

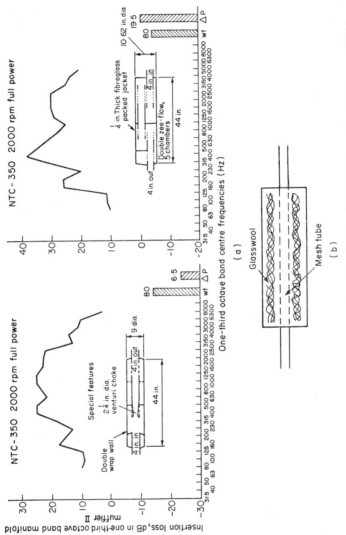

FIG. 4.3(a).   Insertion loss for two types of muffler. (K. Bender, Bolt, Beranek and Newman, 1973.)
(b) High frequency silencer.

Research, Southampton, over more than 15 years, two main conclusions can be drawn for frequencies higher than 500–1000 Hz[1] (Fig. 4.4).

1. For the petrol engine (Otto cycle), engine noise is proportional to the logarithm of engine speed (N). The level of noise can then be expressed as follows (octave band sound pressure level):

$$SPL = 50 \log N + L'_0$$

2. For the diesel engine, engine noise is also proportional to the logarithm of engine speed but the increase with speed is less:

$$SPL = 30 \log N + L_0$$

The relation between the engine speed and the spectra of the noise explains the slope of the spectra for the frequencies above 100 Hz: for a tenfold higher engine speed $10N$ and a diesel engine, the noise level radiated at a definite frequency is 30 dB higher—at a tenfold frequency, the noise level is the same (see Fig. 4.4(b)). The excitation tends to be

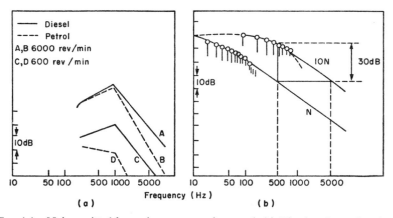

FIG. 4.4.   Noise emitted by engine versus engine speed. (a) Diesel and petrol engine; (b) spectra for diesel engine—engine speed N and $10N$ (from Priede[1]).

constant when speed is increased and when for high frequencies many natural modes are excited. The result is a slope of the spectra for frequencies above f_0 (f_0 is about 500 Hz): −30 dB per tenfold increase of frequency (−30 dB/decade).

For a petrol engine the slope is −50 dB/decade. If dBA is accepted as a unit representing the overall level of noise, the low frequency response (for a diesel engine) is not so important as the response in the frequency from

500 to 1000 Hz. It may therefore be concluded that the noise from the engine and the structure is proportional to the logarithm of the operating speed N:

$$\text{SPL (dBA)} = 30 \log N + \text{L}_0$$

For a light car engine, f_0 is greater and the preceding explanation is not quite correct. However, experimental results are in quite good agreement with the following expression for the noise level:

$$\text{SPL (dBA)} = 50 \log N + \text{L}'_0$$

Engine Load

For the noise radiated by engine and structure there is no effect of load in the case of the diesel engine, but for the petrol engine the effect of load is very marked as is shown in Fig. 4.5.[1] This is due to the throttling of the

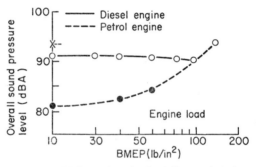

FIG. 4.5.   Noise and engine load (from Priede[1]).

petrol engine intake with no load. Trucks seem noisier when climbing up a slope; the explanation is given on p. 108, but briefly it is that the noise duration is greater and the exhaust noise is greater.

Engine Size

Generally speaking, the engine size is not an operating parameter for the overall sound pressure level. The noise emission is related to the design of the combustion chamber and the weight of the partitions. But for the same design, any increase in the intensity of radiated sound should be due to the increase in size of the radiating surface. Priede found that the overall sound pressure level (dBA) is proportional to the logarithm of engine size and increased by some 17·5 dB/decade with the engine size. This is equally true for petrol and diesel engines, although in practice it is more important for the diesel engine, for the engine size range is greater and trucks, especially, emit engine noise.

Theoretical Expression of Engine and Structure Noises

Since the load has a very small effect on diesel engine noise, the intensity of the overall sound level (dBA) is thus determined by the engine size and by the engine speed as follows:

E.1: $\text{SPL} = 30 \log N + 17 \cdot 5 \log V + L_0$ for diesel engine

E.2: $\text{SPL} = 40 \log N + 17 \cdot 5 \log V + L'_0$ for turbocharged or two-stroke diesel[9]

E.3: $\text{SPL} = 50 \log N + 17 \cdot 5 \log V + L''_0$ for petrol engine (no load)

where N is the engine speed (rev/min) and V is the engine displacement volume; L_0, L'_0, L''_0 are constants.

The engine speed is more important than the engine size. It is well known that a big car driven with a low engine speed is usually very quiet. But at high engine speeds or at maximum accelerations, such as those required by noise certification tests, big cars are noisier. This is, in part, why the ISO norm for the measurement of vehicle noise is undergoing further discussion (see p. 101).

E.3 is not correct for high loads, since load has a marked effect on the noise emitted by petrol engines. A diesel engine works best at a rated speed N which decreases with engine size. We may assume that $N = 2500/(V)^{\frac{1}{2}}$ (Ref. 1); thus the noise level at the rated speed hardly depends on the size. E.1 may be expressed as follows

$$\text{SPL} = 2 \cdot 5 \log V + L_0$$

where V is the engine cylinder volume.

Priede found that a 30 l/cylinder engine running at 500 rev/min develops 6000 hp at the same level of noise dBA as a 0·37 l/cylinder developing only 42 hp at 4000 rev/min. It may therefore be concluded that a good design may enable future truck engines to be quieter by some 10–15 dBA, although it must not be forgotten that future bigger engines will be noisier at low frequencies which are underweighted by the A weighing network (see Fig. 4.2).

Diesel Engine and Petrol Engine

In practice, the noise levels are higher for diesel engines than for petrol engines at lower speeds and lower loads. The injection pumping, the self-ignition mechanism, etc., may produce a noise pressure level 6 dBA higher for a diesel-powered car than for one with a petrol engine.[3] On the other hand, at full load or high rotating speed a petrol engine is noisier than a diesel engine of the same power.

Fan Noise

For well-silenced cars the fan noise becomes important. The fan produces a broad-band aerodynamic noise, and the interaction between moving blades has a marked effect on the noise pressure level. The main parameter to be considered is the tip speed. For a given fan the noise increases by 55–60 dB per tenfold increase of the rotating speed. The blade materials may give slight help in controlling fan noise; cast aluminium blades for example may produce less noise than stamped steel blades.

Practical schemes for reduction of fan noise consist in:

1. Limiting tip speed.
2. Installing a thermostatic clutch.
3. Removing obstructions.
4. Driving fan electrically.
5. Making aerodynamically shaped blades.
6. Reducing inlet turbulence.

A special coupling to reduce fan speed for high engine speeds or for low engine temperature is the most effective way of controlling the noise. For large engines, two fans may emit less noise than a single bigger one.

Tyre–Road Interaction Noise

At speeds ranging from moderate to high (70–150 km/h) the tyre noise predominates. It may even be an important component of the total emitted noise for acoustically well-designed cars and for big American cars at low constant speeds. The tread design and the road texture are the major factors in the production of tyre noise. For the tyre itself, tread wear, speed and carcass design all influence the noise emission.

The experimental tyre noise data are based largely on truck tyre noise measurements made by the National Bureau of Standards and Bolt, Beranek and Newman Inc. in the USA,[4,5] and automobile tyre noise measurements by the Transport and Road Research Laboratory in the

UK,[6] and by Rathé[7] (TRRL also studied truck tyre noise). Pocket-tread tyres produce the highest noise level. Crossbar tyres are quieter than pocket-tread tyres but noisier than rib tyres, as shown in Fig. 4.6. The overall noise level increases by an average of 10 dBA per tenfold increase of speed; the same result has been obtained by Waters[8] and by IRT (France) for car tyres.[9]

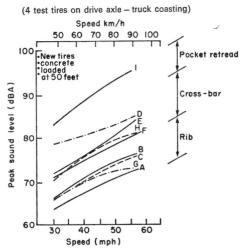

FIG. 4.6.   Lorry tyre noise. Maximum sound level as measured at 50 ft versus speed for a lorry running on a concrete surface. (From Olson, *Highway Research Record*, No. 448, 8, by permission of Ontario Dept. of Highways.)

Theoretically, at least three tyre noise sources have been examined: aerodynamic, air pumping and vibration:[2,5,10]

1. Air pumping is felt to be a major contributor to tyre noise. As a tyre tread segment contacts the road surface, air is squeezed out of small depressions in the road and the tyre interstices. As the tread segment leaves the surface, air rushes back to fill the voids. Hayden[10] has estimated the sound pressure level from pumping air as:

$$SPL = 40 \log S + L_0$$

where S = speed.

The tyre noise increases with the actual width W of the rolling tyre on the road (Waters):

$$SPL = 30 \log W + L_0$$

This explains why a worn tyre may be noisier than a new tyre, although the tyre tread segments are worn down. A large tractor–trailer combination can produce 10–15 dB higher noise levels than a passenger car at the same speed. Little is known about the effect of road surface roughness on the generation of tyre noise.

2. Vibration of the tyre carcass is the second source of noise. At low frequencies below the first carcass mode, the response of the tyre to the road excitation is very low and is likely to be of no consequence. At higher frequencies the carcass responds in a modal manner, although the sound radiated may be little. For still higher frequencies the solid waves in the carcass will radiate very efficiently. On a stone-paved street, the vibration of the tyres may increase the overall sound pressure level by as much as 9 dBA.

3. Aerodynamic noise due to the air flow over the tyre seems unlikely to be a major source, although on a wet surface the sound pressure level increases at frequency ranges above 1000 Hz, so that noise level may increase 7 dBA in rainy weather. The load and the inner tube pressure have only an indirect effect on both the air pumping from the outer sections of the tyre and the width W.

Miscellaneous

The rush of air across the moving vehicle produces more noise inside than outside the cabin, but the relative significance of this effect at very high speeds (more than 130 km/h?) is difficult to estimate.

The vehicle deteriorates with age and, in the case of trucks and buses, with a mileage between 8000 km and 25 000 km or so the noise level may increase by 3–4 dBA.[11]

Little research has been done on gear and brake noise, although these may be important for stop-and-go traffic involving heavy vehicles.

Motor Cycles

For engine volumes larger than 150 cm^3, up-to-date motor cycles emit noise mainly from the engine. The difficulty of silencing small motor cycles is due to cumbersome silencers. According to Olson,[12] in contrast to other motor vehicles the engine is practically the main source of noise from motor cycles in good condition. Throttle setting appears to be a more important parameter than engine speed or road speed. Full-throttle operation results in maximum noise regardless of engine speeds or load or the gear in which the transmission is operating. On the road with throttle setting only sufficient to maintain a steady speed, the noise level increases with road speed—12 dBA per doubling of speed. The noise levels depend largely on the model and very little on the engine displacement (Fig. 4.7). For medium sized engines, the noise levels at full throttle may be in the range 95–77 dBA (77 dBA for BMW, 600 cc, two cylinder, four-stroke). With four-stroke engines the noise level may be lower than with two-stroke engines. Motor cycles used by young drivers often emit more noise from the exhaust, since the drivers deliberately modify the silencer to increase the noise.

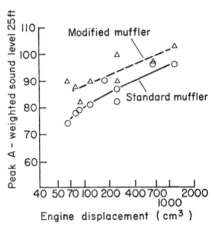

FIG. 4.7. Maximum noise levels of motor cycles versus engine displacement. (From Olson, *Highway Research Record*, No. 448, 10, by permission of Ontario Dept. of Highways.)

Isolated Vehicles in Operation (Cars)

Effect of Speed

The engine and exhaust noise are generally predominant at speeds below 60–70 km/h. The increase of engine, exhaust, inlet and tyre noises as has been mentioned previously means that:

1. The maximum noise level emitted by a car driven at constant speed in a certain gear is proportional to the logarithm of the vehicle speed.

2. A twofold increase in the speed of a vehicle running in a given gear implies an increase of 10 dBA in the maximum noise level.

The extensive measurements made by Rathé *et al.*[7] on 58 different models of European cars confirm these principles and give the dispersion for the level of the different car models. Olson,[12] who studied American car models, found an increase of 10·5 dBA for doubling of the speed (for speed higher than 30 km/h).

But the maximum noise level is not the only important parameter for the degree of annoyance. For a car passing at constant speed S in front of a standing microphone, the equivalent level of mean acoustical energy during a given time $2T$ may be expressed thus (see Appendix, p. 125):

$$\text{E.4:} \quad [\text{Leq}]^{+T}_{-T} = L_{max} + \log\left(\frac{d^2}{Sd_0} \cdot \frac{1}{T}\right) + 10\log\left[\text{Arctg}\,\frac{2ST}{d_0}\right]$$

where L_{max} is the maximum noise level at a distance d_0; d_0 is the actual distance. If $T = 1$ h, ST/d_0 is usually very great and E.4 becomes with $d_0 = d$:

$$\text{E.5:} \quad [\text{Leq}]_H = L_{max} + 10\log\frac{\pi d}{S}$$

As L_{max} increases approximately as $30\log S$, $(\text{Leq})_H$ increases approximately as $20\log S$.

For speeds below 60 km/h cars are driven in various gears and the increase of L_{max} emitted therefore depends less on the speed (Table 4.2).

Effect of Acceleration

For a petrol engine, the noise increases with the load, and with the acceleration of the car. We have some experimental results from Rathé[7] and will show some from IRT (Institut de Recherche des Transports) for a Renault R.16 car[13] (see Table 4.2 and Fig. 4.8) which illustrate the trend of the L_{max} level for different speeds and accelerations.

TABLE 4.2

L_{max} FOR DIFFERENT SPEEDS AND ACCELERATIONS
(CORRECT GEAR RATIOS)

Acceleration (m/s/s)	$\gamma = 0$	$\gamma = 1$	$\gamma = 2$
Speed km/h			
20	62	64	68
40	66	68	71
60	66–68	67–70	72
90	73	73	

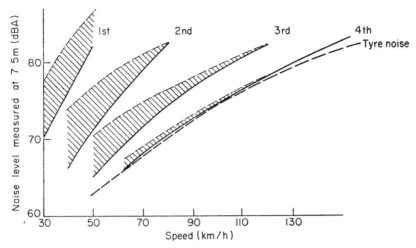

FIG. 4.8. Noise from an R.16 car for different speeds and accelerations—1st, 2nd, 3rd and 4th gear. (From Pachiaudi, *I.R.T.*, C.E.R.N. 1973.)

We see that with correct use of gear ratios, acceleration has more effect on noise level at low speeds, particularly in starting away. The effect of starting away varies and is greater if the cruising speed is reached quickly by maximum acceleration. The effects are illustrated in Table 4.3 (for a Renault R.16[13]).

During varied traffic conditions, the speed and acceleration of a car fluctuate and there is no relation between something like mean L_{max} and, for example, the average speed. One can only say that, for a mobile driver

TABLE 4.3

(a) 0–40 km/h

t.sec	5 sec	5–2–2	4–2–8
γ m/s/s	2·2	1·65–0–1·3	1–0–0·8
t.total	5	9	14
Leq (dBA)	70	64	60

(b) 0–60 km/h

5–2–3·5	5–2–7	4–2–8–2–3·5	4–2–8–2–6·5
2·2–0–1·5	1·65–0–1·3	1–0–1·2–0–0·5	1–0–0·8–0–0·8
10·5	14	19·5	22·5
72	68·5	65·5	63·5

following the measured car, Leq increases proportionally to 20 log S, if S is the harmonic mean speed on a given road for a given car.

Directivity Patterns

As the noise emitted by a vehicle is reflected by the road, the directivity pattern is not spherical. The sound radiation is predominantly along directions 20–40 deg above the horizontal.

Gradient

Figure 4.9 shows the effect of gradient on the noise level emitted by cars; the noise depends mainly on the speed (see Table 4.4 for Leq).

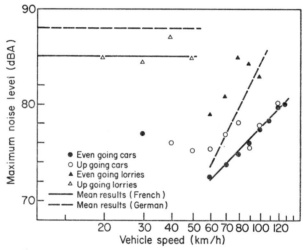

FIG. 4.9.   Noise levels of French vehicles on a 1 in 16 gradient. (*I.R.T.*, C.E.R.N.) and mean German results on a 1 in 14 gradient (Ullrich[14]).

Isolated Vehicles (Diesel Trucks)

Effect of Speed

As has been shown, the engine speed or load increases engine noise rather less abruptly for a diesel engine than for a petrol engine (Fig. 4.10). Moreover, trucks have many gears and more often keep to the rated engine speeds. The result is that the effect of the speed on truck noise is very low when the tyre noise and the exhaust noise do not predominate (a common case for speeds below 60 km/h). For speeds above 50–90 km/h the truck noise, L_{max}, increases 9–10 dB per doubling of the vehicle speed.[12] At low

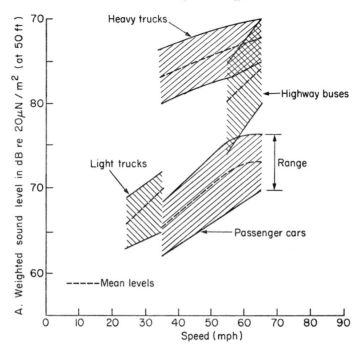

FIG. 4.10.   US single vehicle noise output as a function of vehicle speed. (From Ref. 2 by permission of Dept. of Transport, Washington.)

TABLE 4.4

(from Ullrich[14])

		$S/km/h$	$(\mathrm{Leq})_H$ (dBA)
Truck	Going up	30	51·5
	Going down	79	49
	Horizontal road	80	50·5
Car	Going up	90	41
	Going down	93	40
	Horizontal road	115	43

speed, E.5 is particularly significant for trucks—Leq decreases, theoretically, as the speed increases.

Acceleration Load and Slope

The engine load has no effect on the engine and structure noises. So for a truck with a good silencer, acceleration, load and slope have little effect on L_{max}. Generally, climbing lowers the speed and Leq increases slightly (see Table 4.4 and Fig. 4.9). The figure and table only consider trucks and cars at constant speed on a long slope; at the beginning of the slope, gear shifts may produce higher noise.

Noise Regulations (see also p. 184)

The regulations on vehicle noise emission limit the noise level when tested using the ISO test procedure for moving vehicles, which is virtually identical to the British Standard 3425: 1966, as follows. The vehicle moves along a straight line which coincides with a line 7 m away from the microphone. The sound meter is placed on the same side of the vehicle as its exhaust pipe at a height between 1 and 1·25 m. The vehicle is driven in its lowest gear ratio in such a way that when it is at a right-angle to the microphone, it is at its peak power rev/min and is developing maximum power. The reading to be applied during each test is the maximum noise level indicated by the instrument for a duration of one second (pp. 183–4).

Some countries, Switzerland, Holland and Belgium, add an emission limit for stationary vehicles (for example at peak engine power). The following limits are generally enforced in European countries:

Motor cycles	(dBA)
(a) Not more than 50 cc	73
(b) More than 50 cc but not more than 125 cc	80
(c) More than 125 cc but not more than 500 cc	84
(d) More than 500 cc	86
Passenger cars	82
Light goods vehicle not less than 3·5 tons gross weight	84
Motor tractor not more than 1·5 tons	82

Heavy vehicles	
(a) Of not more than 200 hp	89
(b) Of more than 200 hp	91

Regulations and the Actual Use of Vehicles

The ISO norm defines a vehicle speed which is seldom used in urban trips. It needlessly penalises high-powered cars. Some investigations are in progress to indicate the most frequently annoying speeds, so as to define a test cycle analogous to those used for car pollution. Failing this, one may consider limiting the equivalent level for a small number of standard speeds, the emission for each speed being properly weighted. Thus in France, following observations on urban traffic, UTAC and IRT (Union Technique de l'Automobile et du Cycle, and Institut de Recherche des Transports) are studying a test cycle. This cycle

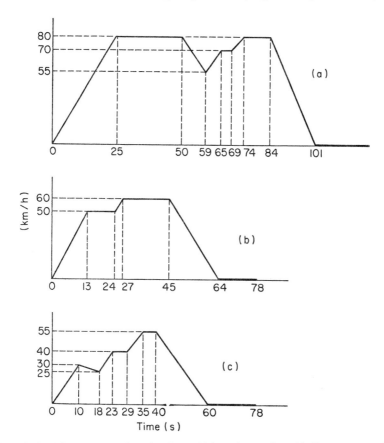

Fig. 4.11. Three proposed cycles for vehicle noise testing. (a) Expressways with junctions; (b) major urban thoroughfare; (c) connecting streets. (From Pachiaudi and Dreyer.[15])

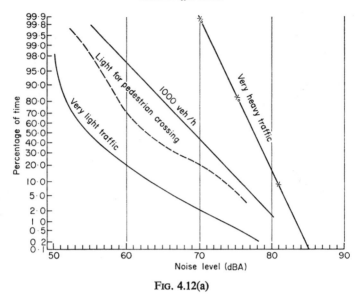

FIG. 4.12(a)

FIG. 4.12.   Noise level distributions on the roadside in different cases. (a) Four types of traffic; (b) at different distances from the road; (c) traffic with 75%, 20% or 9% lorries.

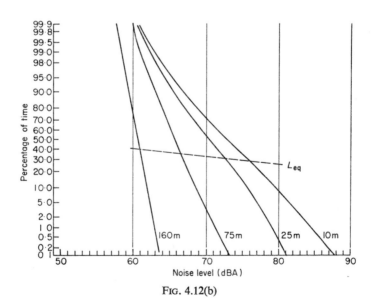

FIG. 4.12(b)

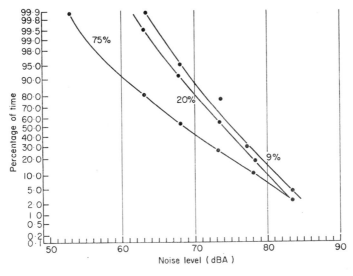

FIG. 4.12(c)

derives in particular from observations on most current ranges of speeds on four categories of road: (1) urban motorways; (2) expressways with junctions (at intervals of 300–1000 m); (3) major urban thoroughfares; (4) connecting streets. The graphs in Fig. 4.11 illustrate these speeds.

NOISE EMITTED BY ROAD TRAFFIC AS A WHOLE

Statistical Distribution of Levels

As we know, the noise emitted by traffic fluctuates, and so long as it could not be represented by a curve of statistical distribution of noise levels over a given period, nothing precise could be said about it. In Europe, the earliest statistical descriptions date back to 1964–66.[16,17]

Each point of the curve of statistical distribution gives the noise level L_t exceeded during a given percentage of time (Fig. 4.12). The scale used on the axis usually corresponds to a Gaussian distribution. The distribution of noise levels in the case of heavy and steady traffic very nearly corresponds with a Gaussian distribution, which is quite convenient.

Indeed, with this scale the distribution curve becomes a straight line which can therefore be defined by two parameters only, for instance the median level L_{50} and the standard deviation of the levels. One has then the following relations, E.6:

$$L_1 = L_{50} + 2{\cdot}35\sigma$$
$$L_{10} = L_{50} + 1{\cdot}3\sigma$$
$$TNI = 4[L_{10} - L_{90}] + L_{90} - 30 = L_{50} + 9{\cdot}1\sigma - 30$$
$$Leq = L_{50} + 0{\cdot}11\sigma^2$$
$$L_{10} = Leq + 1{\cdot}3\sigma - 0{\cdot}11\sigma^2$$
$$L_5 - L_{95} = 3{\cdot}3\sigma$$
$$LNP = Leq + 2{\cdot}56\sigma = L_{50} + 2{\cdot}56\sigma + 0{\cdot}11\sigma^2$$

For heavy traffic (more than 100 v/h) σ at the roadside ranges about 5 (one can often accept that $Leq = L_{50}$, $LNP = L_1$, for in the most common case σ is about 3–4 dB). A little more precisely, $Leq = (L_{50} + L_{10})/2$ and $L_{10} = Leq + 3 = L_{50} + 5$. These relations, E.6, are not exact in the case of non-Gaussian distribution such as: (*a*) very light traffic; (*b*) superimposition of the noise emitted by two roads with distinct traffic streams; (*c*) road intersections. Figure 4.12 represents some cases.

Acoustic parameters used to characterise the annoyance due to the noise are deduced from statistical analyses of the noise level. Peak levels are approximately represented by L_1, if heavy vehicles are absent.

Incidence of Traffic Volume and Traffic Speed

For expressway and free-flowing traffic with few trucks, measures enable a theoretical law giving L_{50} or Leq to be verified, at 3 m from the roadside: E.7: $Leq = 10 \log Q + 20 \log S$ (Q being the traffic volume in veh/h, S the speed in km/h.[16] This law is based on the fact that the noise level emitted by a fast-running vehicle increases as $30 \log S$. It should be noted that the traffic volume is actually the product of the stream speed by the concentration of vehicles on the road. The laws that relate the parameters L_{10}, LNP, TNI to vehicle flow are more complex and one may prefer to use nomograms like those of Fig. 4.13.

These clearly show that L_{10} and consequently L_1 vary with the traffic much less than L_{50} or Leq. σ decreases when the flow becomes heavier, so for LNP and TNI, which are related to σ, the increase is modified by the diminution of σ. One will find in Nelson[18] several equations to calculate precisely L_{10}, L_{50} and L_{90}. When there are no heavy vehicles: $L_{10} = 8 \log Q + 20{\cdot}4 \log S + 13$ (3 m from the roadside).

Formulae and nomographs are not accurate when there is any considerable dispersion of individual vehicle emissions. This occurs in the case of light traffic offering highly dispersed speeds or a considerable percentage of trucks (more than 10–15%). When the traffic becomes rather heavy (more than 600 veh/h for each lane), the increase reduces the stream speed, so that the noise level Leq, L_{50}, L_{10} would reach maximum figures for approximately half the highest traffic volume, which is approximately 1400 veh/h per lane. As for TNI and LNP, their decrease with the rise of the flow to more than 400 veh/h is continuous.

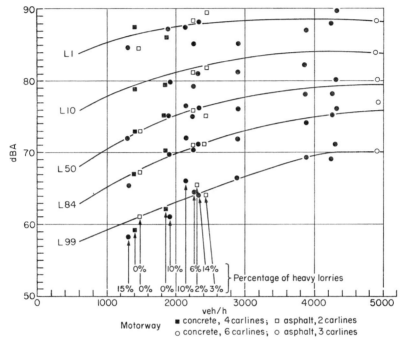

Fig. 4.13.　Noise levels L_1, L_{10}, L_{50}, L_{90}, σ at a distance of 3 m from the motorway (traffic speeds 100–110 km/h.) (From Auzou and Lamure.[16])

Upgrades

We have very little information as regards noise on upgrades. As we saw, the noise emitted by a car depends largely on the engine load, and hence on the slope of the road. At the same speed, the peak levels are higher on a gradient than on a level road. For a 1 in 14 gradient, the peak level and the Leq level increase 4–7 dBA if the speed is maintained around 60 km/h. On a highway the speed may scarcely drop, and the noise levels remain unchanged. Generally, Leq increases greatly as the speed drops (see pp. 108–109).

Heavy Traffic Including Trucks

For traffic which includes a high proportion of trucks, we must take into account the traffic volume. If it is low the presence of trucks does not modify the average speed of the stream. If it is high, incorporation of numerous trucks lowers the traffic speed.

(*a*) *Uniform speed unrelated to the traffic volume.* For a steady stream at uniform speed, the presence of trucks notably increases the noise levels (L_{10}, Leq and LNP). Methods for predicting the noise level in Leq differ according to the research workers and the country, probably because the truck fleets and truck definitions are different (see Fig. 4.14).

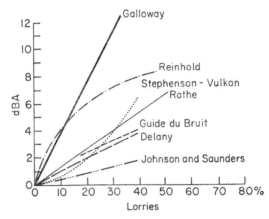

FIG. 4.14.   Additional value for Leq versus various percentage of trucks. Results from Galloway, Reinhold, Stephenson and Vulkan, Rathe, Guide du Bruit, Delany, Johnson and Saunders.

Two simple methods can be used. The simplest one consists in increasing the level obtained with E.7 of:

> 0 dB for 10–15% of trucks
> 2 dB for 20%   of trucks
> 3 dB for 30%   of trucks
> 4 dB for 40%   of trucks
> 5 dB for 50%   of trucks

Another consists in assuming that a truck emits an acoustic power K times higher than vehicles cruising on a level road. K depends on the speed of the cars and of the trucks. K may reach 20 if the speeds are the same, 80 km/h. Formula E.7 may be used with the consequent figure for Q. One may take

$K = 5$ for a moderately congested stream on a level road. Generally $K = 10$ is assumed for an upgrade (see below). For the prediction of L_{10}, the nomographs may be preferred (see Fig. 4.15).

The latest and still unpublished investigations of the IRT and CERN tend to base the calculation of Leq on the noise level of a given traffic

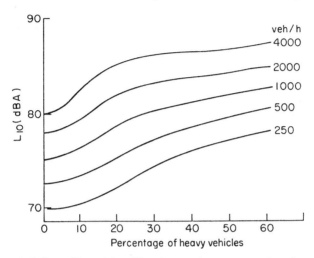

FIG. 4.15. Variations of L_{10} with traffic volume and percentages of trucks, mean speed of 96·5 km/h; at 10 m from the roadside. (From Ref. 19.)

stream without heavy vehicles, to which is added a correction depending on the percentage of lorries and the nature of the traffic. Thus the study shows that with an identical percentage of lorries the increases of Leq are much higher in the case of an urban road with heavy traffic than on a motorway with very low traffic volumes.

TABLE 4.5

Leq CALCULATED BY FAVRE AND IRT
(see Appendix)

Slope	0%								7%							
Traffic volume	1 000				4 000				1 000				4 000			
% Trucks	0	25	50	100	0	25	50	100	0	25	50	100	0	25	50	100
Leq	67	68·8	70	72	72.5	74·3	75·6	76·7	67·3	72·7	75	77·5	73·3	78·8	81	83·5

(*b*) *Speed related to vehicle flow.* When the traffic is rather heavy an increasing percentage of trucks reduces the mean speed, and the noise level increases are all the less important as the total flows are higher. Table 4.5 indicates the levels reached in various conditions. The TRRL gives accurate equations for use with digital computers.[18]

When there is a high percentage of trucks, some workers assume that Leq is almost independent of the mean speed of the traffic (Fig. 4.16).

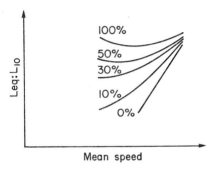

Fig. 4.16.   Variation of Leq with the mean speed and the percentage of trucks (horizontal road).

Upgrades with Heavy Goods Traffic

On a long, little-congested upgrade truck speeds decrease. Their power of noise emission depends very little on the slope or on the speed (see Fig. 4.9 and Table 4.4). Leq becomes a little higher and the K equivalent of a truck for light cars may rise to ten for a 1 in 14 upgrade if we remember that $(Leq)_H = L_m - 10 \log S$. Thus for 20% of trucks and a 1 in 14 gradient the noise level increases and Leq ranges about 4 to 5 dB more than the Leq without trucks. This increase reaches 8 dB only in extreme cases (50% of trucks and a slope of 1 in 14) (see Table 4.5). The peak level and to some extent the increases in L_{10} are less marked. If there are few trucks (less than 10%) the gradient hardly changes the noise levels Leq and L_{10}.

For a long but little congested gradient two comparisons may be made:

1. If the upgrade is weighed against a fast horizontal road where the speeds are rather high, one should remember that the truck speeds may very much decrease in the gradient; consequently, Leq will increase all the more as the speed drops. Reducing the speed to one-half, for example, will increase Leq by 3 dBA, although the peak level will not change.

2. If, in an urban area, the upgrade is represented by a road with a gentle gradient and slow-moving traffic (less than 60 km/h) the result is a slight decrease in speed and a slight increase in noise levels (Leq, L_{10}, L_1).

If the upgrades are rather congested, introducing significant percentages of trucks reduces the general speed of the traffic stream and the increases in noise level due to the gradient are slighter. Table 4.4 which summarises the relationships of traffic flow, composition and gradient indicates the expected increase in Leq. L_1 increases 5 dB while TNI, LNP or σ may actually decrease on an upgrade. To conclude: increases in noise level in short gradients are complex because the changes in gear ratios are frequent and the consequent use of higher engine speeds does not limit the responsibility for high noise levels to trucks only. Because of this defect of short gradients, the substitution of underpasses for conventional junctions proves moderately efficient in reducing noise.

Junctions—Fluctuating Speed

Very few studies have been devoted to noise levels in the vicinity of a crossroads or in congested areas with stop-and-go traffic. Noise level predictions must therefore be made from knowledge of noise levels emitted by individual cars.

Near a crossroads, the noise levels depend very much on the relative positions of the noise-measurement point and the crossroads. The level fluctuations, hence σ, LNP and TNI, rapidly decrease with the distance from the road. On the other hand L_1 and Leq may be comparatively insensitive to the distance downstream from the crossroads, for the noise level increase due to the acceleration of the vehicles has to be compared with that resulting from the speed reduction for trucks (see Table 4.6). Olson[12] found that the noise levels produced by acceleration or at the cruising speed may be equivalent (see also Fig. 4.17).

TABLE 4.6

NOISE LEVELS NEAR A CROSSROADS (ONE-WAY STREET—TWO TESTS, 15 m LONG MEASUREMENTS)[13]

	At the crossroads	*50 m after*	*80 m before*
L_1	82–81	82–81	74–72
L_{10}	77·6–76·5	76·3–76·5	68·8–68·2
Leq	73·1–72·7	72·2–71·3	65·7–65·3
LNP	85·9–84·5	84·4–82·4	74·4–74

The noise levels depend very much on the starting conditions and the intensity of acceleration: Favre and Pachiaudi found a 10 dBA diminution for L_1, L_{10} and Leq, 50 m after the startpoint (a nearby minor road compelled drivers to slow down).[13] Considering our present state of

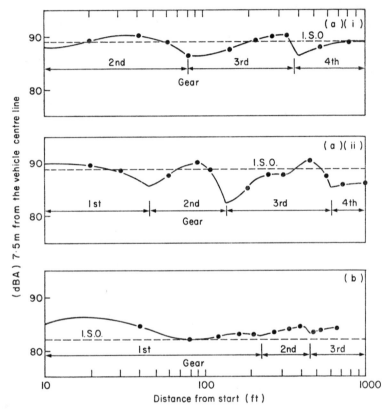

FIG. 4.17. Noise during maximum acceleration. (a) 5·9 litre diesel truck (i) unladen, vehicle weight 8504 lb, maximum acceleration from a standing start in second gear; (ii) fully laden vehicle weight 21000 lb, maximum acceleration from a standing start in first gear. (b) 1·6 litre petrol car; normal laden, maximum acceleration from a standing start in first gear. (From Raff and Perry.[20])

ignorance about the physical laws and the characteristic indices of the annoyance due to the noise in the vicinity of a crossroads, the town planner or the traffic engineer will make due allowance notably for the number of inhabitants located before and after a junction and the rated speeds after the junction (see pp. 168–169). The rated speeds indeed determine the length of the road following a junction which comes under its influence (Fig. 4.17).

The presence of heavy vehicles further increases the level fluctuation due to braking and starting and adds much more to Leq than cars since, in the case of trucks running under 60 km/h, Leq is largely determined by their speed.

A congested street emits a noise level that can hardly be predicted from the volume and speed of the traffic. The effect of the street width can be taken into account (see pp. 169–170). In Lyon, in three cases where stop-and-go congestion was eliminated by means of one-way streets, the noise levels L_1 and Leq have hardly varied. On the other hand, the standard deviations have increased from 3–4 dB to 5–5·5 dB. For two-way streets, and for traffic volume exceeding 1000 veh/h for 3 lanes, one may consider that Leq no longer increases with the traffic volume (for a 15 m wide street Leq then ranges about 74 dB). For a very low traffic volume, Leq increases as $10 \log Q$.

Speeds are increased by the use of one-way streets; as compared with a normal traffic situation in a two-way street, L_{10}, Leq or L_1 will increase about 5 or 7 dB for the same traffic volume. LNP gives a slightly higher rate of increase.

These elements make clear why the use of one-way streets in modern traffic engineering does not help in reducing the noise level.

Traffic Noise Variations Over 24 h

The closely related variations of Leq and L_{10} mainly depend on the traffic volume and they also show a seasonal and weekly periodicity. The study of the variations over 24 h is most important, particularly in defining indices of annoyance.

In a working day, the following characteristics can generally be observed in major built-up areas.[12,21,38]

1. The daytime level between 8 h and 18 h is practically unaffected by peak hour traffic. Its Leq level can be characterised equally well by the Leq from 8 to 18 h, the Leq from 11 to 12 h a.m. or the level of the mean traffic volume over 24 h.

2. The quiet night level is much less distinct. Its duration is highly variable for it can range from 12 to 7 h in medium-sized cities and from 2 to 4 h on some heavy traffic arteries (Figs. 4.18 and 4.19).

   In such a case the difference in Leq between the day level and the offpeak night hours falls to less than 10 dB, whereas it reaches and exceeds 20 dB in big city quiet streets.

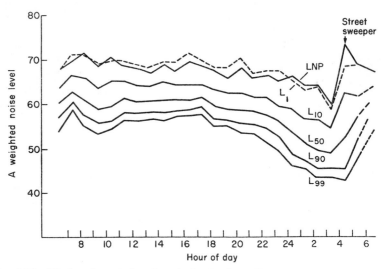

FIG. 4.18. The hourly variations in noise level values of L_1, L_{10}, L_{50}, L_{90} and LNP at a typical commercial area in Medford, US. (From Wesler.[21])

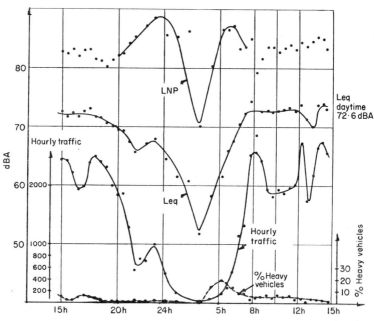

FIG. 4.19. The hourly variations in noise level values of LNP, Leq, traffic volume and percentage of lorries near a motorway in France. (From Pachiaudi and Dreyer.[15])

Some workers, like the authors of the Swedish guide '*Urban Planning and Noise from Road Traffic*' (see p. 124) think that, with due allowance made for this situation and for night comfort requirements, it is the night level which should determine the design of the road system and the traffic management. In fact, the periods of night silence should be characterised mainly by their duration and by the LNP or TNI levels, particularly at the beginning and the end of the night (see Chapters 1 and 2).

3. In the evening, Leq falls rather slowly from 20 to 24 h and even more slowly in the centres of very big cities or where there is mainly heavy vehicle traffic. As against this, LNP would show a somewhat irregular variation. Generally, the peak LNP reached at a particularly unfavourable hour must be retained as a characteristic of night annoyance.

4. At the end of the night the noise increase is much more sudden and Leq may rise 20 dB between 4 h and 7 h. LNP offers as sudden a rise but its maximum is followed by a fall due to the day traffic. On some French thoroughfares, with heavy night traffic, this maximum appears between 5 h and 6 h. Here again, the moment is a particularly unfavourable one.

Naturally, the result of such observations may be different if the traffic is light, for LNP then appears to vary in a capricious way. In small towns having a single industry or in residential areas, marked peaks of traffic determine the noise. On the whole, although the variations of noise levels are related to the traffic, it appears that they have to be examined in a very different manner—peak hours, for instance, offer no particular interest for noise levels.

Propagation of the Noise Emitted by a Road
Theoretical Attenuation by Geometrical Divergence

With increasing distance acoustical energy is dispersed over larger and larger surfaces. For an individual vehicle at a distance d, the sound attenuation ΔL comparative to the level at distance d_0 is:

$$\Delta L = 20 \log \frac{d}{d_0}$$

For L_{50} and Leq near a road (linear source):

$$\Delta L_{50} = \Delta Leq = 10 \log \frac{d}{d_0}$$

and for L_1:

$$\Delta L = 20 \log \frac{d}{d_0}$$

The attenuation of peak levels, or L_1, is then greater than that of L_{50} or Leq[22,33] (Fig. 4.20).

In the same way σ, and to a lesser extent LNP, decrease more rapidly with distance than Leq. The distance from the road must be doubled for Leq to decrease merely 3 dB if there is no other attenuation.

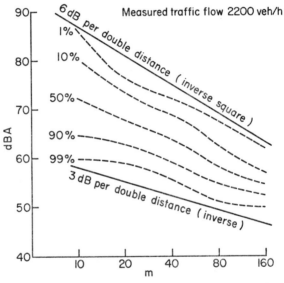

FIG. 4.20. Attenuation of noise with distance for L_1, L_{10}, L_{50}, L_{90}, L_{99}. (From Langdon and Scholes.[33])

When the road is partly hidden, the noise level Leq depends on the 2φ angle under which the carriageway is viewed from the point of measurement (see Fig. 5.3, p. 141). If L_0 is the referential level for an infinite road, the level will become $L_0 - 10 \log \pi/2\varphi$. It can thus be observed that if the visible part of a road is reduced to one quarter, the level decreases 6 dB. But if the reduction of the acoustical energy received theoretically depends on nothing but the 2φ angle, it will in fact be much more important for the distant areas, since air and ground absorption and various diffractions will interfere (see below).

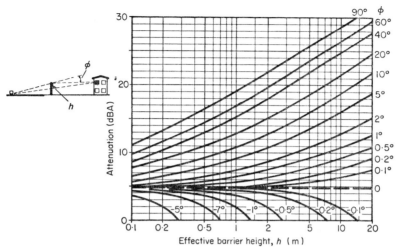

FIG. 4.21. Sound attenuation by a screening wall (point source)—nomogram of Readfearn.[24]

Diffraction and Attenuation by Barriers

Interposing an obstacle in the path of a sound wave causes diffraction and attenuation, the theoretical calculation of which for a point source can be obtained from the nomograms of Readfearn[24] (Fig. 4.21), or Scholes and Sargent (Fig. 4.22) who use the path difference $\delta = a + b - c$ between the diffracted and the direct ray. Readfearn uses the apparent

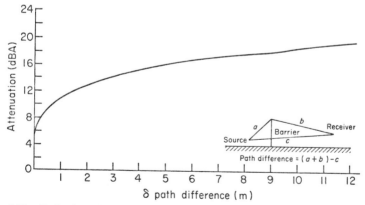

FIG. 4.22. Reduction of road traffic noise L_{10} by a very long barrier—nomogram of Scholes and Sargent 'Designing against noise from road traffic', *Applied Acoustics*, **4**, 3, 203–34 (1971).

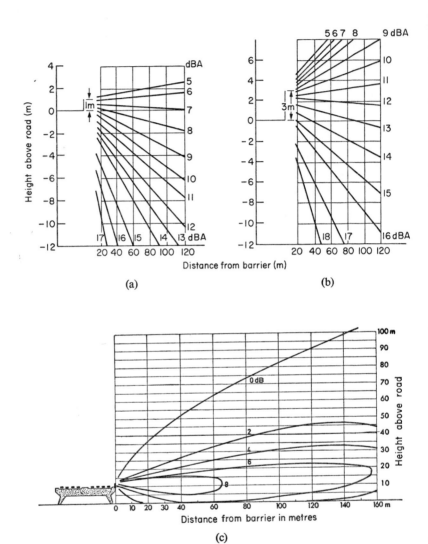

FIG. 4.23. Performance of noise barriers. (a) Reduction of L_{10} by a 1 m high barrier (from Scholes and Sargent[29]); (b) reduction of L_{10}, a 3 m high barrier (from Scholes and Sargent[29]); (c) reduction of L_{50}, Leq by a 2 m high barrier (from Rapin[28]).

height of the barrier and the breaking angle of the diffracted rays. The attenuation increases greatly with the sound frequency and the nomograph used for road traffic tends to confuse the attenuation of the overall levels in dBA and the attenuation at 500 Hz. One will notice that the attenuation is rather variable when the shadow plane is crossed, since it rapidly shifts from 0 dB to 8 dB in the shadow area for an angle of only 1 deg.

For the complex sources constituted by actual roads, precise attenuation calculations are often difficult in practice[26] (see pp. 148–149). The graph of Readfearn is often adequate if the road can be considered as linear, infinite and loaded to capacity. In this case, the attenuations of peak level and Leq due to the diffraction are assumed to be the same. For more precise computations, the Maekawa procedure is often preferable. In practice, one can use either nomographs, computation programs (see pp. 122–124), standard list (see p. 156), buildings as barriers, or various isophons (Fig. 4.23).

Different authors have tested the various noise reduction prediction procedures. The difference between the results may be as high as 6 dB in unfavourable cases, as in the case of very small values of path-length difference δ.[27]

Sound Absorption

Sound absorption by air at long distances is only a few dB per 100 m for frequencies exceeding 1500 Hz (4 dB/100 m at 2000 Hz; 0·5 dB/100 m at 500 Hz). This attenuation, then, matters little on the global level in dBA for car traffic, though it can lower the noise pitch at distance.

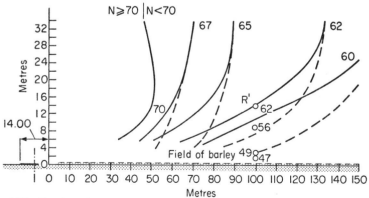

FIG. 4.24. Isophons for sound transmitted near a straight road and a traffic flow of 2000 veh/h. L_{50} A weighted noise level—effect of soft ground, e.g. bare soil. (From Auzou and Lamure.[16]) (– – – bare soil; —— barley field.)

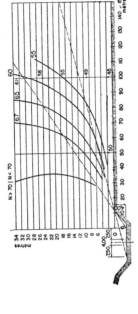

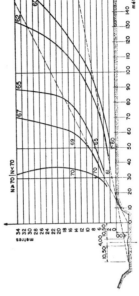

The 60 dBA isophon is displaced about 50 m towards the cutting (traffic without heavy vehicles and ground effects)

A low embankment reduces the expected effect due to the ground

As opposed to the preceding case the ground effect is low, the profile is widely flared and the traffic includes heavy vehicles

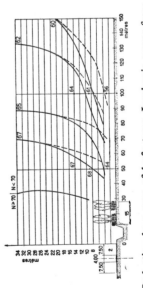

Reduction by a row of leafy trees. Isophonic curves from Fig. 4.24 (bare soil)

FIG. 4.25. L₅₀ isophons for a straight road and a traffic volume of 2000 veh/h—A weighted noise level. (From Auzou and Lamure.[17])

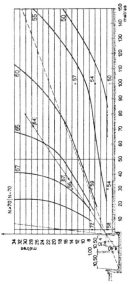

Noise levels in the Montrouge cemetery adjacent to the boulevard peripherique of Paris. In brackets, the estimated levels along the semi-masked area median

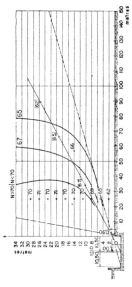

Noise levels adjacent to the boulevard peripherique of Paris (asymmetrical cutting)

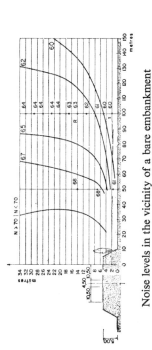

Noise levels in the vicinity of a bare embankment

The height of a viaduct reduces the ground effect (latticed side fences)

Fig. 4.25—contd.

Sound absorption of ground grazing waves over distances exceeding 40 m is a complex phenomenon in which the thermal turbulence and, of course, the nature of the ground are both involved. At a distance of 100 m from a road, one can observe an additional attenuation of more than 8 dBA if the ground is covered with shrubs or high and dense grass. This accounts for the usual slope of isophons in the vicinity of a road[29] (see Fig. 4.24).

Taking due account of its importance for grazing propagation one sometimes adopts a coefficient of 0·4 dBA of additional attenuation per metre when calculating the global noise level at points inside the ground effect area (for another calculation and for a point source, see Fig. 4.26 and Ref. 18). The wind and the night fall of temperature may also considerably raise the noise level at a distance. These differing effects make evaluation of the efficiency of acoustic barriers a delicate task. Figure 4.25 shows isophons in various cases of diffraction or ground effect near a road.

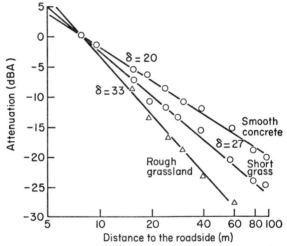

FIG. 4.26. Attenuation with distance characteristics for various ground surfaces (point source). δ denotes the power index defining the attenuation of sound intensity with distance. (From Nelson.[18])

Noise on a Road Flanked by Buildings

The multiple reflections in a street produce a reverberating field whose acoustic level depends very little on the position and height of the point of measurement. For a given traffic the noise level decreases 3 dB for doubling of the street width. In practice, one can adopt for Leq a formula of the following type:

$$\text{Leq} = 15 \log Q - 10 \log l + 38$$

where *l* is the width of the street in metres.[30] The noise increase in a similar situation but without buildings thus ranges from about 6 to 9 dB. It is almost independent of the relation between the height of the houses and the width of the street.

Prediction of Noise Levels

The levels in the vicinity of an unobstructed road may be predicted. When there is no ground absorption, no wind and no thermal turbulence, one can use such theoretical formulae as:

E.8: $\text{Leq} = 52 + 10 \log \dfrac{Q}{d}$ (private cars on motorway without speed limitations but below 120 km/h)

E.9: $\text{Leq} = 10 \log Q + 20 \log S - \delta \log d + L_0$

E.10: $L_{10} = 7 \log Q + 5 \cdot 5 \log S - \delta \log d + 11$ (see Ref. 31)

E.11: $L_{10} = 8 \log Q + 20 \cdot 4 \log S - \delta \log d + 27 \cdot 4$ (see Ref. 18)

S is the mean speed in km/h; *d* is the distance to the middle of the road; $\delta = 10$ for E.9, $\delta = 10 \cdot 7$ for E.10, $\delta = 16$ for E.11. When there is sound absorption by grassed areas the coefficient δ may rise to 18 for Leq.

Within 40 m of the roadside the distance effect is much feebler if the highway is wide (4 traffic lanes or more) and nomograph methods are preferable. The useful distance to the roadway is then the geometric mean of the distance to each traffic lane.

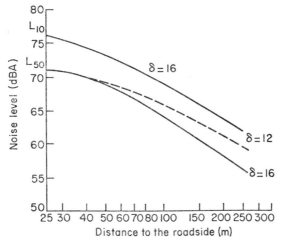

FIG. 4.27.   Practical effect of distance on L_{50}, L_{10} for a traffic volume of 2000 veh/h (4 car lanes).

In practice, if the roadside inhabitants are to be given reasonable predictions of noise levels, particularly in considering the night temperature inversion which upsets the ground absorption, it is wise to adopt a value of about 12 for δ, that is a 4 dBA attenuation of L_{50} or Leq when doubling the distance. For noise peaks or L_1, attenuation of 7 dBA for each doubling of the distance is more appropriate. For L_{10}, $\delta = 16$, the attenuation when doubling the distance is 5 dBA (Fig. 4.27).

In the case of a sound attenuation caused both by geometrical divergence and distance, we have different calculations. For the peak levels the attenuations are calculated for each vehicle separately and merely summed to obtain the overall attenuation. For Leq and L_{50}, it is usual to add the two attenuations for all the vehicles together. But for computation programs each vehicle is considered as a distinct source.

Combination of Noise from Several Traffic Streams

The estimated Leq level is obtained by a simple logarithmic combination, starting from the Leq levels of each roadway. The calculation of L_{10} is more complex and one can use the useful algorithm and computer model developed by Watkins and Nelson. The same model is applicable for LNP and TNI which may decrease with noise superimposition.[34]

Reduced Scale Models

Where there are diffractions and multiple reflections, theoretical calculation of noise levels becomes so complex as to be almost impossible. In such cases, acoustic fields may be simulated on 1/10th or 1/20th scale models using either noise sources at 5000 or 10 000 Hz, or from analogy with electric or electromagnetic fields.[32] In Europe the British NPL and the French CSTB usually employ scaled-down acoustic models. Precise control of air humidity enables a close analogy to be maintained with atmospheric attenuation at normal frequencies. Thus, R. Josse of CSTB is developing a Scale Models Centre at Grenoble.[36] Figure 4.28 shows a typical simulation of a traffic route flanked by buildings.

In conjunction with a computer program simulating the traffic whose fluctuations have to be taken into account, such scale models readily predict the various acoustic parameters desired.

Simulation by Computer

Computer programs enable traffic flows with differing speeds and composition to be simulated, together with attenuation effects from fairly simple geometries with from one to as many as four diffractions or sound

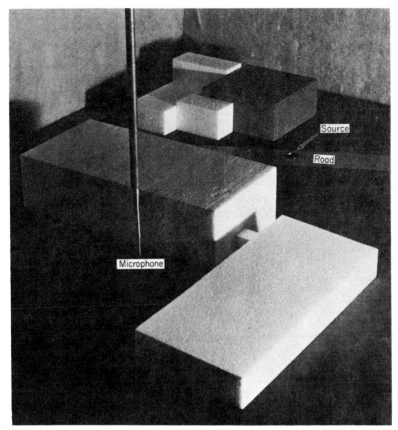

FIG. 4.28

reflections. Such programs are all the more necessary, as the numerous variables discussed above are interdependent, making precise calculation extremely difficult in most cases.

The programs offer generally the following advantages: reproducibility of calculated levels; simulation of varied traffic; calculation of various acoustic parameters or indexes. Reproducibility is particularly necessary if legal actions or infraction of regulations are to be avoided.

Among the existing programs may be mentioned those of Department of Transportation, Washington 1972 (Galloway),[35] Delany,[31] Ström[37] and Nelson.[18,34] The IRT–CERN in Lyon maintains a program which, in particular, should give the different acoustic parameters using the results obtained on scale models by the CSTB in Grenoble. In the USA, many States have their own computer programs. Generally these are issued from the initial Dept. of Transportation program though Bolt, Beranek and Newman and the DOT are now about to prepare a testing procedure for these many different programs.

Useful Guides for Planners

Various countries provide guidelines which aim to enable non-specialists to perform simple calculations of the noise levels produced by the traffic. The following may be cited:

'Urban Planning and Noise from Road Traffic' (in English), National Swedish Board of Urban Planning.

'Noise Assessment Guidelines', US Department of Housing and Urban Development, Bolt, Beranek and Newman, Inc., August 1971.

'Le Guide du Bruit', Ministère de l'Equipement et du Logement, France, 1971.

'Noise Prediction Method', in National Cooperative Highway Research Program, report US 117, digest in Research Results Digest, Digest 24 Dec., 1970.

'Noise Prediction Method', US Department of Transportation, 1972.

British guides: Dickinson, J. 'New Housing and Road Traffic Noise, a Design Guide for Architects', *Design Bulletin* No. 26, DOE, HMSO, 1972; Building Research Station, 'Motorways, Noise and Dwellings', *Digest* 153, HMSO, 1973; DOE, 'Planning and Noise', *Circular* 10/73, 19 Jan. 1973; 'A Design Guide for Engineers', Highway Research Board. Nat. Acad. Sciences, Washington DC, 1971.

CONCLUSION

The prediction of noise levels in the vicinity of carriageways has developed enormously in the past few years. Whereas we knew hardly anything about it as recently as ten years ago, we now possess highly sophisticated devices and techniques such as advanced computer programs. Nevertheless, research must still be actively pursued to enable noise to be predicted from disordered traffic, such as at street intersections, or where traffic volumes are low.

APPENDIX

COMPUTATION OF Leq CONSIDERING TRAFFIC LAWS
(from Favre)

Theoretical Formula

Considering a vehicle to be an acoustic omnidirectional point source moving on a straight road, at a constant speed S, the Leq received by an observer at a distance d from the road is, in first approximation:

$$[\text{Leq}]_{t_1}^{t_2} = L_0 + 10 \log_{10} \frac{d_0^2}{sd} - 10 \log_{10} (t_2 - t_1) + 10 \log_{10} \left(\text{Arctg} \frac{vt}{d}\right)_{t_1}^{t_2}$$

where L_0 = sound pressure level emitted by the vehicle at a distance d_0, t_1 and t_2 times such that, at time $t_0 = 0$, the vehicle passes by the observer (Fig. 4.A1).

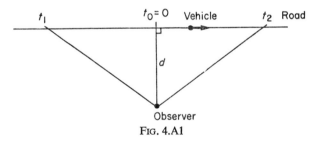

FIG. 4.A1

For a traffic of Q vehicles per hour with the same acoustical and dynamical characteristics,

$$\text{Leq} \sim L_0 + 10 \log_{10} Q - 10 \log_{10} v - 10 \log_{10} d - 7{\cdot}4 \text{ (dBA)} \quad \text{(A.1)}$$

with L_0 in dBA ($7{\cdot}5$ m/vehicle); s in km/h; and d in m.

We make the assumption that the observer is not too far from the road, so that the ground effect and molecular absorption are negligible.

For several classes of vehicles, say light vehicles and heavy vehicles, the total Leq will be the combination of $(\text{Leq})_{\text{LV}}$ and $(\text{Leq})_{\text{HV}}$.

125

Traffic Laws

For a straight road, there is a relation between several traffic parameters such as volume, speed, composition (percentage of heavy vehicles, for instance), slope, number and width of lanes, type of road and speed limits. In France, following studies performed by SETRA, analytical relations and nomographs have been published which allow a fair prediction of these parameters (MATELT–DRCR 'Calculs de Rentabilité Appliqués aux Investissements Routiers', Manuel d'application, 25 June, 1970).

TABLE 4.A1

EVALUATION OF V_{LV} AND V_{HV} FOR A GIVEN SET OF ROADWAY AND TRAFFIC DATA

Traffic volume (veh/h)	% HV	2 × 2 motorway— no speed limit		2 × 2 road— no speed limit	
		V_{LV} (km/h)	V_{HV} (km/h)	V_{LV} (km/h)	V_{HV} (km/h)
1 000	0	113	—	98	—
	25	107	75	95	72
	50	103	75	85	63
4 000	0	100	—	80	—
	25	95	72	saturation	
	50	91	70	saturation	

Therefore, for a given set of roadway data we can estimate the value of V_{LV} and V_{HV} (mean speed of light and heavy vehicles, respectively) as a function of Q (traffic volume) and p (percentage of heavy vehicles).

For instance, we can see on Table 4.A1 an evaluation of V_{LV} and V_{HV} for the following roadway and traffic data: 2 × 2 lane width motorway and

TABLE 4.A2

MEAN MAXIMUM SOUND PRESSURE LEVEL EMITTED BY SINGLE VEHICLES CRUISING AT A DISTANCE OF 7·5 m FROM THE OBSERVER

Road with slope 0%	Speed (km/h)	Maximum level at 7·5 m (dBA)
Light vehicles	120	81
	100	79
	80	76
Heavy vehicles	80	84
	60	81

2 × 2 lane width road; slope 0%; 1000 and 4000 vehicles per hour; 0–25% and 50% heavy vehicles.

Noise Emission of Single Vehicles

For computing Leq with formula (A.1), we have to know the mean sound level received at a distance d_0 from the different vehicles cruising at a speed V. Such results were obtained by measurement of single vehicles. For instance, with $d_0 = 7\cdot5$ m, and for a smooth asphalt surface, we obtained the results shown in Table 4.A2.

Computation of Leq

Using these values, it is now possible to compute Leq. The results obtained correlate well with the measurements. They show that the effect on the noise emitted by traffic of parameters such as percentage of heavy vehicles or speed depends on other traffic or roadway parameters. The computations are summarised in Table 4.A3.

TABLE 4.A3

VALUE OF Leq FOR DIFFERENT ROAD AND TRAFFIC CONFIGURATIONS: DISTANCE 30 M FROM TRAFFIC STREAM; SURFACE, SMOOTH ASPHALT

Q (veh/h)		1 000			4 000		
% HV		0	25	50	0	25	50
2 × 2 motorway	Leq LV	67·2	65·7	63·8	72·5	70·4	68·7
	Leq HV	—	66·0	69·0	—	72·0	74·6
	Leq total	67·2	68·8	70·2	72·5	74·3	75·6
2 × 2 road	Leq LV	65·7	64·4	62·5	70·5		
	Leq HV	—	66·0	67·5	—		
	Leq total	65·7	68·3	68·5	70·5		

REFERENCES

1. Priede, T. Origins of automotive vehicle noise, *J.S.V.*, **15** (1), 1971.
2. *Motor Vehicle–Highway Systems Noise*, Serendipity, Arlington, Virginia, 1970.
3. Bruckmayer, F. 'The Problems of Fixing and Gradually Reducing the Maximum Permissible Levels of Noise from Motor Vehicles', Meeting of International Association against Noise, 1964.

4. 'Peak A Weighted Sound Levels due to Truck Tyres', (a) US Report OST-ONA 71, Dept. of Transportation, Washington DC, 9 Sept., 1970; (b) Report N, OST-TST 72, National Bureau of Standards (Office of Vehicle Systems Branch), 1 July, 1972.
5. Leasure, W. A. and Bender, E. K. Tyre–road interaction noise, Proceedings, Internoise 1973.
6. Underwood, M. C. P. 'A Preliminary Investigation into Lorry Tyre Noise', Department of the Environment, TRRL, Report LR. 601, 1973, Crowthorne (Transport and Road Research Laboratory).
7. Rathe, E. J., Casula, F., Hartwig, H. and Mallet, H. Survey of the exterior noise of some passenger cars, *J.S.V.*, **29** (4), 1973.
8. Waters, P. E. The control of road noise by vehicle operation, *J.S.V.*, **13** (4), 1970.
9. Favre, B. 'Le Bruit des Pneumatiques', IRT Report, 1974.
10. Hayden, R. E. 'Roadside Noise from the Interaction of a Rolling Tyre with the Road Surface', Proceedings of the Purdue Noise Control Conference, Purdue University, Lafayette, Indiana, 1971.
11. 'Urban Traffic Noise—Strategy for an Improved Environment', O.E.C.D., Paris, 1971.
12. Olson, N. Survey of motor vehicle noise, *J.A.S.A.*, **52** (5), Pt. 1, Nov. 1972.
13. Favre, B. and Pachiaudi, G. 'Emission Sonore d'un Vehicule en Accélération —Application au Bruit de Carrefour', 8th International Congress on Acoustics, London, 1974.
14. Ullrich, S. Fahrgeräuschpegel von Pkw und Lkw an einer 7% igen Steigungs und Gefällstrecke, *Kampf dem Lärm*, **4**, 1973.
15. Pachiaudi, G. and Dreyer, P. 'Proposition de Cycle d'essai pour Caractériser le Bruit Émis par un Véhicule. Etude de Comportement d'un Véhicule en parcours Réel', IRT, France, April 1973.
16. Auzou, S. and Lamure, C. 'Les Niveaux de Bruit au Voisinage des Autoroutes Dégagées', Cahier du CSTB, No. 599, December 1964.
17. Auzou, S. and Lamure, C. 'Le Bruit aux Abords des Autoroutes', Cahier du CSTB, No. 669, February 1966.
18. Nelson, P. M. 'A Computer Model for Determining the Temporal Distribution of Noise from Road Traffic', Department of the Environment, TRRL Laboratory Report 611, 1973, Crowthorne (Transport and Road Research Laboratory).
19. 'Etude des Coûts Sociaux des Transports Routiers Urbains', 18e Table ronde d'Economie des Transports, Conférence européenne des Ministres des Transports, April 1972.
20. Raff, J. A. and Perry, R. D. H. A review of vehicle noise studies carried out at the Institute of Sound and Vibration Research with a reference to some recent research on petrol engine noise, *J.S.V.*, **28** (3), 8 June, 1973.
21. Wesler, J. E. Community noise survey of Medford, *J. Acoustical Society of America*, **54** (4), October 1973.
22. Rathe, E. J. Note on two common problems of sound propagation, *J.S.V.*, **10** (3), 1969.
23. Scholes, W. E. and Sargent, J. W. 'Motorway Noise and Dwellings', B.R.S. Digest 135, HMSO, London, 1971.

24. Readfearn, S. W. Some acoustical source observer problems, *Philosophical Magazine and Journal of Science*, T.200, p. 223, 1940.
25. (a) Maekawa, Z. E. Noise reduction by screens, *Mem. Fac. Eng., Kob Univ.*, 11, 29, 1965 and 12, 1, 1966; (b) Noise reduction by screens, *Applied Acoustics*, 1, 1968.
26. Kurze, U. J. and Anderson, G. S. Sound attenuations by barriers, *Applied Acoustics*, 4, 1971.
27. Kugler, B. A. and Piersol, A. G. 'Field Evaluation of Traffic Noise Reduction Measures', BBN–Highway Research, Record No. 448, 1973.
28. Rapin, J. M. 'Etude des Modes de Protection Phonique aux Abords des Voies Rapides Urbaines', Cahiers du CSTB, No. 952, May 1970.
29. Scholes, W. E. and Sargent, J. W. Ground effect and motorway noise, *Applied Acoustics*, 5, 1972.
30. Josse, R. *Notions d'Acoustique*, Editions Eyrolles, Paris, 1972.
31. Delany, M. E. 'A Practical Scheme for Predicting Noise Levels L_{10} Arising from Road Traffic', NPL Acoustics Report AC 57, July 1972.
32. Lyon, R. H. 'Propagation of Transportation Noise', Proceedings, Internoise 1973.
33. Langdon, F. J. and Scholes, W. E. The traffic noise index; a method of controlling noise nuisance, *Arch. J.*, 147, April 1968.
34. Nelson, P. M. The combination of noise from separate time varying sources, TRRL, *Applied Acoustics*, 6, 1–21, 1973.
35. Galloway, W., Clark, W. and Kerrick, J. 'Highway Noise. Measurement, Simulation and Mixed Reactions', Highway Research Board Report 78, 1969.
36. Rapin, J. M. 'Mise au Point et Première Application d'une Méthode d'Étude sur Modèle Réduit de la Propagation des Bruits du Trafic Routier', Cahiers du CSTB, No. 810, August 1968.
37. Ström. EDB Program for prognosering av veitrafikkstør, Teknisk Rapport L.B.A. 549, Technical University, Trondheim, Norway.
38. Buchta, E. 'Uber Phänomene die bei der Bestimmung des erforderlichen Schallschutzes für Anlieger an geplanten Verkehrsstrassen zu beachten sind', Kampf dem Lärm, February 1973.
39. Muster, O. and Brammer, A. J. 'Noise Radiation from Engine and Drive Train', Proceedings, Internoise 1973.

CHAPTER 5

METHODS OF TRAFFIC NOISE REDUCTION

C. LAMURE

VEHICLE IMPROVEMENT—BUSES AS AN EXAMPLE

Substitution of good public transport systems for private cars is plainly helpful for reducing road congestion, air pollution and energy consumption. On the other hand, the effectiveness of public transport as a means of reducing traffic noise is less obvious. Provided they are not extremely noisy, public transport vehicles can fairly easily bring about a reduction in mean noise levels or, more precisely, in Leq, i.e. the energy mean noise level over a specified period, at peak hours, but vehicles of very high acoustic quality are needed to obviate the higher peak levels which are particularly annoying when traffic is light, or at night, or in pedestrian zones.

Let us consider the present outlook for reducing the noise of buses, for they will still account for the main bulk of public surface transport for a long time to come.

BUS NOISE

Noise levels from present-day buses unfortunately far exceed those from cars and are thus a separately identified environmental nuisance. For example, when 1400 persons an hour are carried in 40 buses instead of 1200 cars, the average noise level Leq falls by at least 5 dBA as compared to the noise the cars would make, but noise peaks will be from 8 to 10 dBA higher.

As a general rule, of course, buses are not entirely substituted for cars; however, promotion of bus transport leads to increased noise unless car traffic is either completely banned or more tightly packed as a result of providing reserved lanes for buses. When, in extreme cases, it is intended to eliminate light vehicular traffic entirely on a carriageway in order to make the utmost use of its capacity for bus traffic, noise levels rise very steeply, i.e. by about 5 dBA for Leq and 10 dBA for peak noise. When traffic is light, or at night, buses in motion or stopping and starting cause far more noise pollution at present than cars, but somewhat less than heavy trucks.

Reserved lanes for public transport services in principle lead to higher top speeds for the vehicles using them, and hence to higher noise emission from such vehicles. This effect is usually limited in daytime because traffic even on reserved lanes is impeded by other vehicles and there is a high level of ambient noise, but peak noise can still be somewhat higher—from 5 to 8 dBA in the worst cases; for instance, in the case of a contra-flow reserved lane alongside dense slow-moving traffic (e.g. the Rue de Rivoli in Paris at certain hours). The increase is slighter during the evening off-peak hours, as reserved lanes then have little bearing on vehicle speeds.

The stresses borne by reserved carriageways sometimes make it necessary to strengthen road surfacings, and in unfortunate cases, where noisy material such as stone paving is used, the noise level can be substantially higher when vehicles are running at high speed. The conclusion to be drawn from all this is that better facilities for bus traffic flow or promotion of public transport are clearly insufficient and that the vehicles themselves must also be improved. A reduction of 10 dBA in the levels of noise emitted by most buses in use today would bring bus noise down to the same level as for cars.

Sound levels in our cities would then be directly proportional to the logarithm of traffic flow and transfers from private to public transport would have a direct bearing on improvements to the environment. There would be a distinct advantage in cases where bus services were running in streets banned to private cars in accordance with the practice which is beginning to be adopted in the UK (Oxford Street, for example).

The Control of Bus Noise Emission by Regulation

The European regulations for new buses permit noise emission in the region of 90 dBA under ISO test conditions, i.e. roughly 7 dBA higher than for private cars.

The following are guidelines set by the EEC for buses and coaches of the maximum permitted weights shown:

Under 3·5 tonnes	84 dBA
Over 3·5 tonnes	
engines of under 200 hp	89 dBA
engines of over 200 hp	91 dBA

The ISO standard test conditions based on a speed of 50 km/h at full throttle (or more precisely at three-quarters of the engine speed at which the engine develops maximum power in a particular gear) correspond to the maximum level that a vehicle may emit in urban conditions, but do not take into account noise emitted when starting off, or with the engine idling, whereas these 'speeds' are very frequent for buses in city traffic. Given satisfactory traffic conditions, it may be estimated that from 15 to 25% of the operating time of a vehicle is accounted for by waiting periods in traffic and at bus stops or intersections.

These standard test conditions can, however, be used for comparing the acoustic quality of different types of conventional diesel buses, since the levels of noise emitted at various engine speeds in their case are proportional to the ISO levels. Table 5.1 gives some levels for the least noisy buses, the others generally attaining 88–90 dBA.

The mean level obtained during type approval tests is in the region of 88 dBA, most readings being in the region of 86–90 dBA.

TABLE 5.1

TYPE APPROVAL TEST MEASUREMENTS FOR PARTICULARLY
QUIET COACHES AND BUSES

1971	Kassbohrer coach SETRA S80(551) model	83 dBA
1972	Bressel BL 55S/040	83 dBA
	Leyland Engine cyl 6540 cc	
1973	Mercedes Benz 0305-407 model	79 dBA

Diesel Buses

A very large majority of European buses have diesel engines. Their running costs are low and they have a very long life, at least as regards the body—sometimes over 40 years. Life-spans such as this are now being shortened by rising repair and maintenance costs, but still account to some extent for the narrowness of the market and hence for the sluggishness of technical progress recorded up to a few years ago.

Campaigns for a better environment have recently given a fillip to research and it has been shown that the noise emitted by a bus could be brought down to the same levels as for cars. The techniques adopted for diesel vehicles are fairly conventional and are partly described in Refs. 1 and 2. Sound insulation of the engine compartment, and modifications to air intake systems and exhaust silencers, all raise problems of space and cooling, but these can be solved. Thus treatment of the engine compartment of London buses already achieved a reduction of 6 dBA more than six years ago.

Sound insulation problems are indeed more easily resolved when enough space is available. For instance, with flat engines housed on the side at a fair height above ground level, there is sufficient room for a sound-absorbing chamber for the air intake and for a sound-insulating panel under the engine.

This is the solution adopted by Mercedes Benz for their 0305 and 0307 buses, which are now being marketed with an ISO sound level about 12 dBA below that of the standard type buses. The noise level when idling is only 55 dBA which means that the vehicle is well-nigh inaudible in an urban environment.

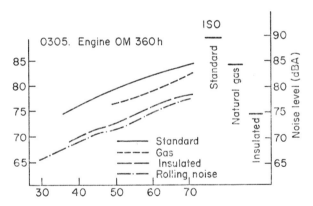

FIG. 5.1. Noise emitted by Mercedes 0305 buses.

In sum, the only noise emitted by the 0305 bus at steady speeds, between 40 and 70 km/h, is vehicle rolling noise[3,4] (see Fig. 5.1 and Table 5.2).

Other makers have also obtained good results. Examples are Magirus Deutz with their Vov M. 170-SH-110 bus which emits from 5 to 6 dBA less than it did before acoustic treatment; Saab Scania with their CR-111; and the double-floored Metro Scania built jointly with a British maker,

TABLE 5.2

NOISE LEVELS (dBA) FOR VARIOUS EUROPEAN BUSES

Vehicle	Cruising speed		ISO Standard
	At 7·5 m	At 15 m	
Standard diesel Average 45–50 km/h	78	72	89
Diesel after acoustic treatment 45–50 km/h	70	67	77
Standard diesel (600 rev/min) Stationary vehicle (2 600 rev/min)	64 82		
Diesel (Mercedes) after acoustic treatment (600 rev/min) Stationary vehicle (2 600 rev/min)	52 74		
MAN Elektrobus 45 km/h	77		77
MAN Elektrobus Stationary vehicle	60		
Trolleybus 20 km/h	70		

i.e. Metro Cammel. These last two emit from 75 to 77 dBA under ISO standard test conditions, i.e. from 8 to 10 dBA less than the ISO standard bus.[4] The internal noise level is thus 68 dBA. Acoustic treatment in this case is based on the same principles as those adopted by Mercedes for the 0305; enclosure of the engine called for a better cooling system with two radiators and two thermostat controlled fans.

British Leyland have brought out in 1974 their new 'super-quiet' model which had an external noise level of 76 dBA when trials were conducted early in 1973. Turbocharging can reduce noise levels on heavy vehicles by 6 dBA but, unfortunately, it is hardly warrantable for engines under 200 hp. However, it may be useful to mention the R.11.14 (141 hp) Ford coach which has more elaborate sound insulation.

The rear-engine layout now generally adopted on one-man city buses is rather useful from the acoustical point of view, as it dispenses with the long exhaust pipe and transmission system which used to radiate considerable noise. The engine can be positioned well up without spoiling the driver's field of vision, as was the case with front-engined vehicles, and there is enough ground clearance for a sound-insulating panel under the engine.

All these good results involve only slightly increased costs: roughly from 2 to 4% of the selling price. The additional costs for Saab-Scania, Mercedes and Magirus Deutz are from Fr. 4000 to 6000 for vehicles within the Fr. 150 000 to 200 000 price range.

As regards research on diesel-engined vehicles in the USA, attention may be drawn to the Mitre Corporation's work on a 1963 GMC 53-passenger vehicle fitted with a catalytic exhaust silencer and driven by a V6 engine.[2] Acoustic improvements were directed to enclosure of the engine and to low exhaust and inlet frequencies (special resonators being fitted at these points). It appears that lack of space prevented acoustic treatment of the air intake from being carried to such lengths as on the Mercedes. Under test conditions fairly close to the ISO standard the noise level fell from 84 to about 77 dBA.

In 1973, the US Government financed the construction of nine prototype non-polluting urban buses by General Motors, A.M., General Corporation and Rohr. General Motors has adopted gas turbine propulsion for one of its prototypes, whereas the other prototypes are to have diesel engines. These buses are to be capable of a speed of 110 km/h.

It is thus possible with well-designed diesel engines to produce very quiet vehicles whose sound level, except when stationary or starting off, is practically entirely due to vehicle rolling noise. Regrettably, there are a few drawbacks: smoke and smell and, failing highly elaborate treatment, noise when idling. The City of Basle authorities do not allow engines to be kept running when the vehicle is stationary. Pollution of all kinds when the vehicle is stopped or starting off is avoided by using compressed-air starters.

Gas-fuelled Buses

While there is no question of using gasoline-engined buses in Europe, petroleum gases provide a cleaner form of bus propulsion than diesel fuel, besides being somewhat less noisy especially under ISO standard test conditions and with the engine idling, and given equivalent acoustic treatment (see Fig. 5.1, where levels of different types of Mercedes 0305 are compared). It may be noted that the use of liquefied gas should help to deal with the additional cooling and fan noise difficulties often encountered in connection with acoustic treatment; it would suffice to find a suitable way of recovering the 'cold' from the expansion valve.

Among other things, liquefied gas makes it possible to use a different type of fuel without any major modification of the engine; all that need be done is to substitute an ignition system and mixer for the standard diesel injection system. The only obstacle lies in providing for safe storage and distribution.

The Air Liquide Company, for instance, has recently developed a filling plant which might be acceptable for use by transport workers.

Saviem and Gaz de France are working on the conversion of a six cylinder engined bus from diesel to liquefied gas, while British Leyland, and Steyr in Austria, are introducing gas-fuelled buses.

The Steyr 'City-Bus', more particularly intended for low-speed traffic in pedestrian zones, is fuelled with liquefied propane or butane. It is a minibus with capacity for 10 seated and 10 standing passengers and is capable of a speed of 40 km/h with 20 passengers. At 10 km/h it emits 64 dBA at 7 m, that is, a lower noise level than that encountered in pedestrian zones. It complies with the pollution control requirements laid down in California. The City of Vienna has ordered 150 buses of this type.

The MAN Company has converted its 750-HO-M10 model for use with a 'non-polluting' natural gas engine, and Mercedes has adopted compressed gas for its 0305 experimental bus.

The development of buses fuelled with liquefied gas seems unlikely to be much influenced by changes in fuel prices and availability as gas and oil prices are linked and as fuel costs at present account for less than 10% of total transport cost.

For immediate purposes, buses fuelled with liquefied gas may be limited to city centre zones or streets, pedestrian zones, transport services for trade fairs and exhibitions, airports and spas.

Hydrogen fuelling need be mentioned only for the record as it still seems a remote prospect and as it is not yet known whether the corresponding noise levels will be an attractive proposition.

New Heat Engines

Research and development on heat engines using various cycles has received a strong impulse in recent years, especially because they can reasonably be expected to be particularly clean. External burners for the steam engine and the Stirling engine, and the fact that neither of these engines nor the gas turbine involve any explosion, are reasons for acoustics engineers pinning their hopes on these new systems. The problem here usually lies in the acoustic treatment of the very large fans needed to obtain an acceptable Carnot efficiency, as the fuel consumption of such vehicles is regrettably high.

The State of California, which has been showing close interest in new systems for clean-engined buses, has launched an experimental steam bus ('Rankine cycle') project.

The noise levels recorded for the experimental projects listed in Table 5.3 are still fairly high, but their scatter makes it clear that useful progress is possible. The noise level now claimed by the Lear Company for its steam bus is substantially lower.

TABLE 5.3

COMPARATIVE EXTERNAL NOISE LEVELS FOR STEAM AND DIESEL BUSES IN THE USA
(range 15 m)[4]

	Steam			Diesel			
	Brobeck	Lear	PS	AC	STA	MUNI	SCRTD
A. Full throttle acceleration 40–60 km/h	76	85	80·5	78·5	84	86	82
B. Full throttle standing start	74	88	86	88	89	90	94·5
C. Idling	68·5	78	78	75·5	78	74·5	75·5

Some European companies, for example, Philips, United Stirling (Sweden) and MAN, are considering the use of the Stirling cycle which has been developed for marine engines and seems rather promising for large units.

The gas turbine being developed by MAN for engines over 200 hp has a high fuel consumption and the theoretical merits of turbines for reducing external noise levels are not as evident as in the case of the Rankine or Stirling engines. However, as already pointed out, General Motors are working on a turbine bus under a US Government contract.

In short, widespread use of such engines for buses is technically feasible, but has more impact on air pollution control than on noise abatement. Though good results have been obtained with diesel engines—well comparable with those obtained with 'non-explosion' engines—it should be borne in mind that the latter might have more long-lasting acoustic qualities and even lower noise levels when stationary. In these last two respects, however, electric vehicles still offer the most promise.

Electric Buses

Electric traction should, in principle, limit the noise level to that due to vehicle rolling noise, and entirely eliminate noise and pollutant emissions—and even energy consumption—when the vehicle is stationary. It may be recalled that a bus is stationary from 15 to 25% of its operating time.

Buses with conventional lead batteries do not involve any considerable technical problems, but they are costly, their range of action is slight and they need a separate petroleum-fired heating system (e.g. by catalytic combustion or radiant panels). They meet requirements reasonably well in cases where bus services are concentrated to cater for steep traffic peaks, as batteries can then be recharged or changed at off-peak hours.

The electric buses for 50 passengers (23 seated), produced by a French company (Sovel), have a maximum speed of 60 km/h and a range of about 100 km under average traffic conditions. The 4 tonnes of batteries can be changed in a few minutes. These buses were put into commision a short while ago in seven new towns in France (three in each town). In addition, electric minibuses (Sovel AS 9B) with rechargeable batteries have been tested for the last two years by Electricité de France in various French towns (Isle d'Abeau, Clermont-Ferrand, Evry, Saint-Quentin).

TABLE 5.4

NOISE OF A GERMAN ELECTRIC BUS

		Elektrobus	Standard bus
At 45 km/h	right-hand side	76·5	85·5
Top gear	left-hand side	77	85·5
At 45 km/h neutral		77	
Vehicle stationary, fan and compressor still running		60	87·5
Engine braking at steady 45 km/h	right-hand side	78	85
	left-hand side	77·5	89

In Germany, the 'Elektrobus', developed by MAN and Bosch and Varta some years ago, is derived from the standard MAN bus for 33 seated and 66 standing passengers. It can reach a speed of 60 km/h and climb 11% gradients. Its 4 tonnes of batteries are carried on a quickly changeable trailer. Its noise emission performance has been carefully measured at a range of 7·5 m. The results are shown in Table 5.4.[3]

These measurements show that the noise emitted at 45 km/h is much the same as the rolling noise in neutral, which seems high, and that the noise level when the vehicle is stationary is comparable to that of the improved diesel bus.

Pending improvements to battery performance as regards storage capacity and power-to-weight ratio, attention is being given to hybrid vehicles, MAN having given up the idea of marketing their Elektrobus as passenger transport operators found it clumsy to operate in practice.

A hybrid arrangement consists in combining a diesel engine running at constant speed with a buffer battery of sufficient capacity to provide the power needed for speed or acceleration peaks or to run on batteries alone in 'sensitive zones'. The advantages as regards noise abatement are evident when running in such zones. Unfortunately, as these are defined by reference to pollution, the fact that noise will be just as much of a nuisance, if not more, in residential suburbs may be overlooked. Some doubts may therefore still be felt about this hybrid approach unless it is combined with particularly effective acoustic treatment of the diesel engine. However, the chances are that the cost of the vehicle would ultimately be very high. The various hybrids in existence include the OE 302 made by Mercedes, Bosch and Varta which was displayed at Frankfurt in 1969 (a diesel engine of only 65 hp is coupled with a generator and a buffer battery). The Regie Autonome des Transports Parisiens also has a Saviem experimental hybrid bus, used mainly for research on electric transmission, which emits about 80 dBA under ISO standard test conditions. Crompton Leyland Electricars Ltd. are also experimenting with two 26-passenger electric minibuses with a range of 35 miles in urban traffic and a speed of 25 km/h.

Regrettably, trolleybuses disappeared some years ago from most of the cities which had them. They were disliked by town planners and by various road users such as firemen, etc. Of the 55 German trolleybus systems existing in 1955, only nine were left in 1970.[3] The last trolleybus in Britain ran at Bradford on 20 March, 1972. In Switzerland, on the other hand, 18 systems still remain and are even expanding; Schaffhausen, for example, introduced its trolleybus system in 1965. In Berne, the local population rejected the purchase of new diesel-powered buses and voted for the extension of trolleybus services.

The merits of trolleybuses as regards the environment, apart from considerations of visual intrusion, are equivalent to those of electric buses and the noise levels for old vehicles in common use without special features are about 10 dBA under those for diesel-engined vehicles. Noise levels on gradients are very noticeably lower.

Relative costs and availability of electricity might well reinvigorate trolleybus transport in other countries besides Switzerland, but it is hard to imagine the wholesale revival of trolleybuses on public thoroughfares as the initial cost and maintenance of overhead lines, the road space taken up by vehicles and the operating difficulties are somewhat daunting. What can be hoped for, on the other hand, is the development of dual-mode systems where own-track operation with mains supply would be coupled with battery operation on urban roads. Research on these lines is going on in various countries, notably in the USA and in France (by Renault's consultancy subsidiary SERI, which is hoping to have this system adopted for the new town at Evry).

With the US Metrobus system, vehicles can be coupled together on segregated track, and so cater for mass transit roughly equivalent to that of the underground. At exits from segregated track they then split off to their respective destinations. This combines the advantages of bus flexibility and trainload capacity.

With regard to hybrids, reference must also be made to the inertia flywheels used on 'Gyrobuses' in Switzerland some years ago. With the advent of high-performance carbon and boron fibres, far better energy storage capacities and power-to-weight ratios would be feasible than with present-day lead batteries.

The development of hybrid systems of this type has not yet begun and research is, for the time being, confined to flywheels, flywheel bearings and housings. Such research is mainly conducted by Lockheed in the USA. It is also considered in some quarters that storage of 'spring' energy by compression of gases or liquids could compete with electric battery storage. Regenerative braking means that energy need be stored for no more than about ten stops-and-starts.

HIGHWAY DESIGN

Sunken Roads and Sound Diffraction

As was outlined in pp. 115–117, the propagation of noise is notably impeded by diffraction. Building a barrier along the road or sinking it in a cutting provides an additional attenuation ranging from 8 to 15 dBA for Leq and from 8 to 25 dBA for L_1 or LNP, when no portion of the road is seen from the location to be protected and when there is no unfavourable reflection. A very feeble diffraction is sufficient to reach 8 dB. This is why the road engineer can effectively protect buildings by preventing them directly overlooking the traffic. But if any part of the road is visible, there will be only slight additional attenuation, e.g. if the whole of either carriageway is seen, the attenuation will be only about 3 dB (Fig. 5.2). The presence of a large building with a facade parallel to the road and close to it is also likely to upset the effect of diffraction. Figure 5.2 illustrates these two cases. It thus becomes quite difficult to achieve significant attenuation near wide thoroughfares passing across districts with high building densities.

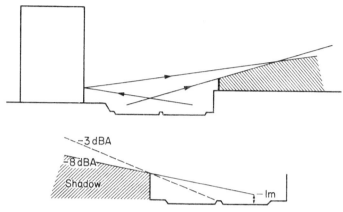

FIG. 5.2.   Sound diffraction and reflection.

Precise calculations of the attenuation due to diffraction can be performed with nomographs or other methods (see p. 115). Figure 5.3 gives the attenuation for two lines of a 2 × 2 road. It is to be noted that in many cases precise calculations are not necessary; for instance, whenever the diffracted rays graze the ground (a frequent occurrence for cuttings in

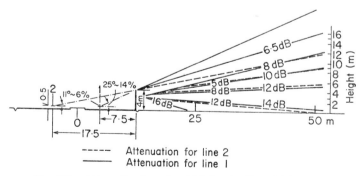

----- Attenuation for line 2
——— Attenuation for line 1

FIG. 5.3.   Attenuation by a 4 m high barrier (from Fleischer[10]).

natural sites and for low rise houses) the ground effect is very important and adds to the attenuation of the diffraction. We would like to quote here Jonasson:[6] 'The great problem is not the acoustic shielding itself, but rather variations of the acoustic properties of the ground and the influence of different weather conditions. Over large distances (70–250 m) a depressed road is more efficient than a barrier with a height which equals the depth of the road. This is due to the fact that the ground attenuation is larger for the low secondary sound source of the wedge. At short distances however, the depressed road is less effective than a corresponding barrier.'

Cuts, Trenches and Road Cover

In the case of trenches in natural soil, with banks covered with vegetation, one may easily reach the basic theoretical attenuation (from 8 to 15 dBA) mentioned for the shadow area; such trenches have obvious attractions in outlying urban areas and areas of low rise buildings—all the more so as they eliminate visual intrusion by the highway. Unfortunately, where high buildings are situated near the road the required depth of the trench may be prohibitive. It may even be geometrically impossible to locate the building in the shadow zone, since an increasing cut may result in the exposure of the building if the natural slopes of the banks are to be retained. Consequently, in the rather critical situations of dense areas it is generally necessary to build road trenches with vertical walls, unless there is enough space to place an earth bank or an acoustic barrier on top of the slopes.

The Bundestrasse B.12 near Munich[7] provides an interesting example of the various possible solutions. To protect the dwellings along that road (mainly two-storeyed houses) the authorities successively: (*a*) lowered the road level and built a wall hidden by coniferous trees; (*b*) covered the road over a distance of 100 m; (*c*) compacted earth banks on top of the slopes, the necessary space being supplied by the housing estate to be protected; (*d*) built up earth banks with low walls on top and lowered the level of the road opposite the five- or six-storey buildings to be protected. However, following this a builder erected two buildings of eight and nine storeys close to the road.

Sunken Cut with Vertical Walls

If the roads are not too wide and the buildings not too near, it is possible to screen the flanking buildings without requiring too much depth. But the reflecting side surfaces formed by the supporting walls may then eliminate the shadow effect. One may have to use absorbent wall surfaces to avoid these reflections. Unfortunately efficient absorbents which also resist bad weather are scarce and expensive. They must offer suitable acoustic qualities and resist rain, solar radiation, dust and oil spray. One possible solution is employed on a current scheme in Berlin: the surface of the wall is formed of concrete blocks with large horizontal perforations backed by a quilt of fibreglass or mineral wool. These must be protected with thin plastic films, although such films are eventually destroyed by solar radiation if not protected from it. If absorbent materials cannot be used, inclined facets reflecting the sound upwards may be employed. The facets must be at least 1 ft square.

Cantilever Screens

When the housing density is very high, when there are visual requirements or when local traffic is to be allowed to continue on the surface, construction of cantilever screens may reduce the noise levels in some highly critical cases (Fig. 5.4).

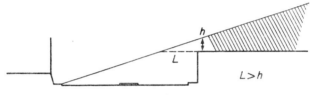

FIG. 5.4.   Cantilever barrier.

This solution is naturally more expensive than vertical walls for a similar sound diffraction. Suitable acoustic treatment of the inner walls of the trench is essential, otherwise the inner reverberation is not only annoying for the car drivers, but turns the sky of the trench into a nearly omnidirectional source of great power, particularly for frequencies below 500 Hz. The acoustic treatment of the inner space may be completed by the construction of midline absorbent walls whose function is not only to reduce the noise level in the trenches but also to act as a screen; this screen also lessens the importance of the cantilevers. Figure 5.5 illustrates a case examined in Neuilly near Paris.[8]

Road Cover

The most radical solution is to cover the road. Cover which is more or less perforated is generally of little acoustic efficiency except for very high

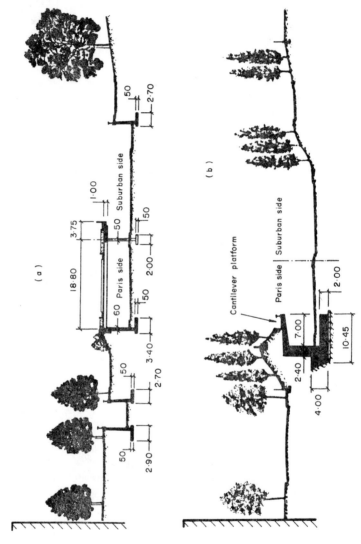

FIG. 5.5. Noise protections in Neuilly: (*a*) road covered over half width (section); (*b*) corbelled cantilever with planted enbankment, giving protection from noise (from Ozanne[8]).

frequencies. Its resistance to wind and snow, as well as the need for surface protections for children's safety, sets difficult problems. Experiments by Rucker and Gluck[9] on tunnel-shaped perforated screens proved disappointing initially, since the results were appreciable only for frequencies above 2000 Hz. Yet the advantage of such devices in respect of cost, visual transparency and ventilation of the road has led Rucker to continue his researches.

The most efficient solution consists in complete roofing or tunnelling; the noise in the tunnel and the noise emitted by the tunnel at each end and from each ventilation station require acoustic treatment using the same techniques as room acoustics. Absorptive surfaces should at least cover the vault; unfortunately, they are expensive and difficult to keep clean and, in consequence, are rarely used. Doubling the absorption area or the coefficient of absorption reduces the level of reverberated noise by 3 dB. Complete separation of the two lanes in a tunnel yields a slight reduction of 2–3 dB with walls of identical quality.

The end of the tunnel constitutes a source with more or less low directivity and may produce a quite considerable noise nuisance, for the emergence of a vehicle causes a sudden rise in noise level. The intensification of the acoustic absorption at the tunnel exit and the use of screens can reduce noise levels at least 10 dB. Without absorbents at the head of a tunnel the noise emitted by the vault is considerable and makes screening difficult. The noise of ventilating plant is induced by the transmission of the inner noise of the tunnel as well as by the aerodynamic noise of the fans. The noise is often excessive at night but the acoustic treatment of the ventilation shafts over several metres should bring about a reduction of the emitted noise to less than 60 dB at 20 m. For a further reduction it will be recalled that the fan noise increases very greatly with the tip speed; reducing the rotation speed, especially at night, or increasing the number of fans are efficient solutions.

In towns, underpasses are more often used than tunnels. Underpasses may improve the environment of existing junctions because they eliminate the start-and-stop at a junction. But multiple short underpasses may be noisier than a long flyover. Underpasses tend to encourage speed because of the initial downward gradient, although heavy vehicles still have to change gear to emerge, so increasing their noise. The width-to-depth proportion of underpasses is generally such that the dwellings nearly opposite are directly affected as has been shown for sunken cuts with vertical walls. The acoustic treatment must therefore be the same as for the sunken cut.

Cost Considerations

The cost of a sunken cut is highly variable and depends on the nature of the ground, the road network to be shifted, etc. Unless other advantages than the reduction of the noise are to be gained, such as the elimination of visual intrusion or a new and more convenient system of local traffic, the cost is often prohibitive and can only be justified in very dense areas where the recovery of available surfaces for pedestrians, parking places or even roads must be taken into account. The cost of concrete flooring (ranging about 700 FF/m² in France) is comparable to the cost of the land or the acoustic insulation of facades. For densities of 40 dwellings/ha or so, with an insulation cost of 5000–8000 Fr. or £500–£800/dwelling, the road cover will rarely be justified in new towns, although for densities of 80 dwellings/ha or more it may be. Generally speaking, strong protective measures can be taken in two instances:

1. When the density of ground occupation is such as to be able to absorb the cost of the entire covering of the road.
2. When the buildings have only one or two storeys, which makes shallow road-trenches or reasonably high screens completely effective.

In other cases, isolated buildings can be protected by specially adapted devices such as very high barriers.

Acoustic Barriers

The function of acoustic barriers is to diffract the sound waves by interfering with their propagation along a straight line. They may be very efficient when they are as high as the buildings or when they are constituted by buildings. In the following paragraphs, however, we shall comment upon acoustic barriers related to the road. The Figs. 4.22 and 4.23 show the theoretically attainable attenuation when there are no harmful reflections and no ground absorption.

Practical Efficiency of Acoustic Barriers

Earth banks or airproof walls may offer more advantages than sunken roads in many cases. Their cost is generally lower and, if necessary, they may be erected after the road has been constructed. But as will be seen they are often not so efficient. The efficiency of full and heavy barriers may be calculated by various methods (see pages 114–115).

In complex cases where multiple reflections are to be expected, measurements on scale models may be preferable. Yet it should be noted that too precise calculations are generally illusory and it would be better to aim at an attenuation of 8–10 dB when locating the areas to be protected in the shadow of the barrier. If 10 dB is the theoretical attenuation 100 m away from the road, the field performance of a noise barrier may vary from 3 dB to as much as 17 dB.[10] Again, a small change of the acoustic admittance of the ground may cause violent changes in the propagation pattern.[11] The additional attenuation obtained from a barrier often falls short of the theoretical prediction: a barrier placed on a site covered with vegetation over some scores of metres will replace the attenuation due to the ground by another not necessarily much greater.[12] For instance, a low barrier situated along a road at ground level some 50 m away from one or two-storey houses will not be of much use if the ground is covered with grass and shrubs to a depth of 50 m. But if the level of the road is notably higher than the surrounding terrain the barrier again proves quite efficient. The ground effect is associated with the effects of wind and of thermal turbulence. The wind caused by atmospheric turbulence can partly impair the efficiency of barriers, and when temperature inversion occurs (notably at night) the sound waves are reversed downward. These effects are not quantified but they may nevertheless be very important. The attenuation of the sound resulting from the barrier may also be impaired by unfavourable

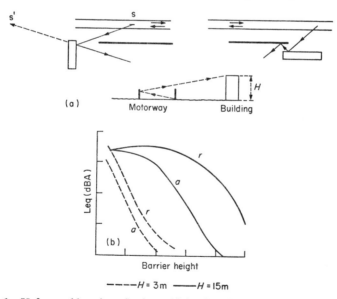

FIG. 5.6. Unfavourable noise reflections: (*a*) barrier efficiency is destroyed; (*b*) noise barriers facing each other—Leq versus barrier height, *a* absorbent barrier, and *r* reflecting barrier (from Reinhold and Burger[13]).

reflections due either to a nearby building or to a bridge, to a wall or to another barrier parallel or perpendicular to the road (Fig. 5.6). We can illustrate the perturbation of the diffraction by analogy with optical reflection of the traffic stream. The importance of the perturbation depends on the length of the traffic stream situated in the field of the mirror (Fig. 5.6(a)). The harmful effect may be eliminated if the reflecting element can be covered with absorbents or constructed with facets oriented in harmless directions; for example, sending the reflected waves upwards.

The attenuations are calculated for the theoretical sources corresponding mainly to the noise emitted from the wheels by cars running rather fast. These sources are 0·5 or 1 m above the road surface. When the traffic contains a high proportion of trucks the attenuation due to a barrier is lessened and can decrease. Thus if the traffic includes 50% of heavy vehicles it can decrease from 10 to a mere 5 dB. The American guide[43] estimates the height of the source will be taken as 2·5 m for a truck.

Length and Mass of a Barrier

A barrier of finite length may not obscure the whole length of the road. The acoustic energy received from the parts not obscured is then proportional to the angle under which the road is viewed from the point of reception. This is exact only if other effects, such as the ground effect or diffuse diffractions along the path of the soundwaves, are neglected, although these are quite considerable. Disregarding them increases the likelihood of calculating a safe minimum attenuation.

If we term Leq the equivalent level of noise received without a barrier, $\Delta\Lambda$ the theoretical efficiency of the infinite barrier and ΔL the minimum attenuation required: the Leq $- \Delta L$ required at the reception point will result from the combination of

$$\text{Leq} - 10\log\frac{\pi}{\pi - 2\varphi} \quad \text{and} \quad \text{Leq} - \Delta\Lambda - 10\log\frac{\pi}{2\varphi}$$

Figure 5.7 gives, for the desired attenuation of ΔL, the necessary theoretical attenuation ΔL related to the angle φ.

One can see that large angles of φ are required—at least 81 deg—for a theoretical efficiency of 8 dB to obtain on average a mere 5 dB as an effective attenuation. This pessimistic calculation for Leq indicates the need for long barriers—6 times as long as the distance between the barrier and the point of reception. If this is impossible, the efficiency of the barrier will be low in reducing Leq but appreciable in reducing L_1 or LNP. Kurze cites the example of a barrier with a theoretical efficiency of -15 dB but with $2\varphi = 60$ deg. The reduction of the sound level will be 4·5 dB for Leq and 7 dB for LNP, because σ becomes $\sigma/(2)^{\frac{1}{2}}$ (Ref. 16). The earthbank at Nienberge in Germany is also an interesting example. At 100 m from the road the attenuation of Leq is 16·5 dBA, 20 for the peak noise level (L_1) and 26 for LNP. The reduction of L_1 depends little on the length of the barrier and is much greater than that of Leq. Moreover

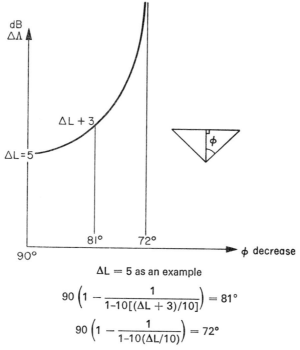

$\Delta L = 5$ as an example

$$90\left(1 - \frac{1}{1-10[(\Delta L + 3)/10]}\right) = 81°$$

$$90\left(1 - \frac{1}{1-10(\Delta L/10)}\right) = 72°$$

Fig. 5.7.   Efficiency of finite length barrier (from Dreyer[15]).

Reinhold[14] of the Bundesanstalt für Strassenwesen at Cologne demonstrated that the reduction of LNP and L_1 are almost identical if $L_5 - L_{95}$ = 10 dBA without the barrier—a frequent occurrence.

The acoustic energy which passes through the barrier without reflection or diffraction is determined by its mass per surface unit. Figure 5.8 indicates the necessary minimum surface mass for the energy transmitted to equal that diffracted. If the barrier is a complex one with two or more walls, its mass may be as low as 5 kg/m². [13] For conditions corresponding to one point of the curve of Fig. 5.8, the attenuation due to a barrier is inferior by 3–5 dB to the theoretical attenuation due to the diffraction. For effective heights which do not exceed 3 m, a surface mass of 15 kg/m² is sufficient to reach the theoretical attenuation. This surface mass, moderate on the whole, allows the use of various materials including wood-panels or plaster-covered wire netting. Obviously the mass and stiffness of the barrier must be sufficient to prevent bending or buckling by high winds.

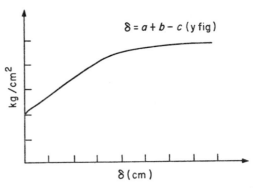

FIG. 5.8.   Equal sound transmission through the barrier and over the barrier (from Kurze and Anderson[16]).

Position of the Barriers

When the surface is horizontal, a barrier attains its highest efficiency when it is as close as possible to either the source of the noise or the point of reception. This is clearly shown with the nomograph. Generally, barriers are placed as near the road as possible, with due regard to the safety requirements of the traffic. Safety and noise barriers must not be confused, although some double-purpose barriers are being studied in the USA[17] and in France, by IRT (see Fig. 5.9). Since such barriers are not yet generally available, the acoustic barrier is generally placed 1·5 or 2 m behind the safety barrier, for in the event of a collision, the displacement of the rail will not exceed 1·50 m. The acoustic barrier is thus generally far from the opposite side of the road. The low efficiency of barriers situated 20–40 m from the opposite roadside is perhaps one of the explanations of the poor results achieved in some cases. The interesting experimentation set out in Ref. 18 and described in Fig. 5.10 for Highway 401 in Ontario was conducted by Harmelink and Hajek. With rather long barriers 3 m high constructed of various materials the attenuation obtained 1·20 m above the ground in front of the houses to be protected ranged from only 2 to 6 dB.

FIG. 5.9

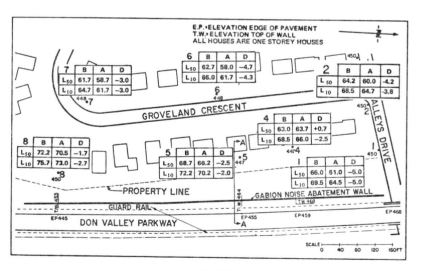

FIG. 5.10(a)

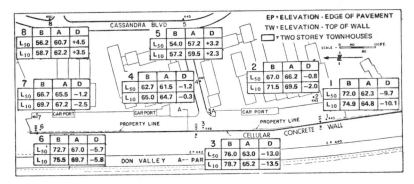

FIG. 5.10(b)

FIG. 5.10. Sound levels before and after construction of (*a*) gabion wall, 9 a.m. to noon and (*b*) cellular concrete wall, 4.15 to 6.30 p.m. (from Harmelink and Hajek[18]).

It may therefore seem preferable to build barriers close to the points to be protected in order to reduce their length and increase their efficiency. Yet this is only effective in areas of low density and where the noise emanates from a point source. One can also use buildings, garages, walls surrounding patios, etc. One method of placing the barrier closer to the stream of vehicles consists in building a barrier between the two carriageways. One can thus build barriers of reasonable height for at least the same efficiency, provided the central barrier is sound-absorbent (Fig. 5.11).

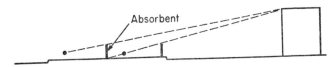

FIG. 5.11. Barrier in the middle of the road—the barrier may be lower in height.

When the site is encumbered with buildings or where there is a complex topography, the choice of the location of the barrier requires much care. Its efficiency will be increased by utilising a building or taking advantage of accidental circumstances of the site, so as to provide a combined protective screen. Figure 5.12 shows one experimental barrier combined with the two embankments of rail and roadways. The gap under the railway bridge was closed off by returning the barrier.

In the case of a road on a hill or in a valley, protection is difficult and the barrier must be more often placed near the building to be protected.

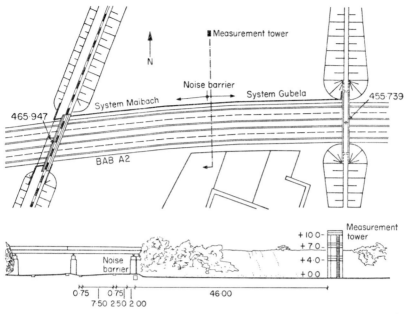

FIG. 5.12. Experimental barrier in Bottrop, Germany (from Reinhold and Burger[13]).

Practical Use of Barriers

The height of the acoustic barrier needed to shield the upper storeys of high buildings close to the road soon becomes prohibitive and visually unacceptable. Whereas the elevated roads on embankment or flyovers diffuse noise more widely than ground-levelled ones, the acoustic efficiency of solid barriers or parapets is better when they are built for such highways, for the cut-off angle for the sound is increased (Fig. 5.13).

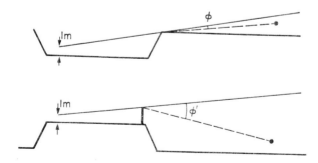

FIG. 5.13. To increase the cut-off angle for the sound.

The flyovers as normally constructed now have solid parapets about 1 m high. For effective results barriers need to be somewhat higher, say 1·5–2 m, assuming that the source is not much above the road surface. Perfecting low barriers and mass producing them would lessen the cost and also reduce visual intrusion. The situation of people living in one-, two- or three-storeyed buildings close to roads, whether elevated or at ground level, could thus be greatly improved. The construction of such barriers for a cost of about £50 a metre would be far more economical than insulating dwellings even for as low a density as 10 houses/ha, which is about the normal private housing density. Rather high walls for shielding high buildings are costlier and involve prejudicial visual obstacles. The most economical and visually acceptable barrier is an earth bank planted with greenery but such methods require a great deal of space.

Design of the Barrier

Unless absorbent or perforated barriers are required, barriers or walls need not have special acoustic properties. In the case where the barrier might reflect the sound towards the buildings on the opposite side, it is more economical to deal with the problem by means of properly oriented facets than with absorbent materials (the requirements concerning these materials have been set out on pp. 146–150). In practice, the main requirements will concern their airproof qualities, their surface mass, their resistance to the wind, rain, snow and fire and their aesthetic appearance.

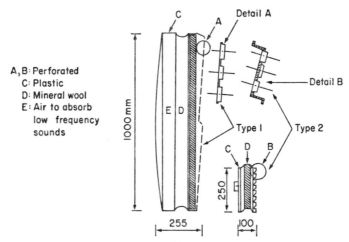

A, B: Perforated
C: Plastic
D: Mineral wool
E: Air to absorb low frequency sounds

Fig. 5.14. Lightweight barrier with absorbent and plastic (from Reinhold and Burger[13]).

They must be easily cleaned and long barriers should be provided with access doors.

Barriers made in sections are preferable to continuous walls for they settle with less trouble and are easier to repair. Nevertheless, barriers framed around wire netting along the motorway and covered with plaster have been suggested in the USA. Sectional barriers may be formed of wood panels or prefabricated brick panels up to 3 m high and 4·8 m long (cf. barriers in Sacramento, USA). In Germany, an experimental barrier has been built consisting of two walls of plastic with mineral wool sandwiched between them. In this case the surface mass of 5 kg/m^2 may be very low (Fig. 5.14). Comparative information about the characteristics and the cost of various solutions of barriers are given in Table 5.5, some others will be found in Ref. 19.

Perforated barriers offer some advantages for ventilation, for keeping clear of snow and for visual transparency. Visual transparency achieved through glass barriers would be very expensive, particularly to keep clean and to repair.

Three Types of Perforation
1. Perforated elements, like a slatted shutter, for use in covers or in tunnels. Those such as were studied by Rucker, which incorporate absorbents, have been mentioned on p. 145. It would seem that the design aims of these devices have not yet been entirely achieved.[9]

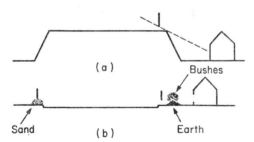

FIG. 5.15.   Perforated barrier.

2. Barriers set up 10 or 20 cm above the ground make fixing and snow clearing easier in some cases. They may consist of solid plates grafted on safety rails. Studies performed by the CSTB, for the IRT and by Reinhold[20] show a loss of efficiency as compared with a full barrier ranging from 1 to 9 dB. This loss may be lower and more acceptable if the barrier stands on planted ground or near slightly banked ground (Fig. 5.15).

TABLE 5.5

ACOUSTIC BARRIERS

Location	Description	Dimension of the barrier (height, length)	Approximate cost per linear metre in dollars
Highway 401 Ontario	Precast concrete wall 10 cm—concrete H columns supports 25 ft apart	8 ft 6 in–11 ft high 2021 ft long	150
Highway 401 Ontario	Precast cellular concrete wall 10 cm 2 steel columns 20 ft apart	8 ft 6 in 800 ft	135
Highway 401 Ontario	Precast cellular concrete wall steel columns 3 m apart	9 ft 680 ft	160
Highway 401 Ontario	Earth Berm	9 ft 10 in 1010 ft	75
Highway 401 Ontario	Precast cellular concrete wall on top of earth berm 2 m high, steel columns 3 m apart	1 m–2·2 m 480 m	100
Don Valley Parkway Toronto	Aluminium wall Aluminium H columns 5 m apart	2·4 m 220 m	120
Don Valley Parkway Toronto	Wooden wall treated fire plywood panels 3/4 in	3 m 120 m	45
Don Valley Parkway Toronto	Gabion wall 36 in	2·4 m 240 m	180
Don Valley Parkway Toronto	Porex concrete wall Steel columns 3 m apart	3·6 m 420 m	90
Motorway No. 4 London	Plastic panel with a metallic frame	2·6 m high 300 m long	100
Sacramento US	Prefabricated brick panels height—1·8 m wide—3·6 m Steel columns		
Hay-les-Roses A.6, near Paris	Wall of concrete— 9 m high		
Bundesautobahn 2 near Bottrop (Federal motorway 2)	Plastic panels 25 cm high 1 m wide 10 cm thick (5 cm mineral wool) Steel column	4 m 200 m	
US	Converting the standard chain link fence into a stucco wall. (Spraying two coats of concrete plaster on each side of the structure.)		

One German case was particularly unfavourable, as may be seen from Fig. 5.15(a). Heaping sand at the foot of the barrier improves its acoustic efficiency while allowing water to run freely.

3. Barriers with successive vertical grooves may give the driver the benefit of visual transparency. Such experimental barriers have been studied in Phoenix, Arizona by the Engineering Corporation of America, using the principle of the movies (rapid succession of images) to achieve visual transparency. The acoustic attenuation is obtained through the vertical prisms which constitute the barrier functioning as Helmholtz resonators.[21] The constituents of the barriers can be of any material without any absorbents (concrete, steel, stucco or wood). An attenuation of 10 dB is obtained for frequencies above 800 Hz. An improvement of the attenuation for low frequencies is still to be sought for. It is possible that visual transparency might be obtained by simpler devices such as walls with minute perforations.

It is evident that useful investigations may be carried out to improve the acoustic protection of people living along important motorways. But a general improvement in the traffic may also be beneficial to those living along secondary roads, as we shall see.

NOISE ABATEMENT THROUGH URBAN PLANNING

No clear-cut method exists for defining specific noise-abatement targets in urban areas, nor are there yet any reliable indicators describing the annoyance caused by traffic noise or for evaluating the effects of some particular form of urban planning.

It should, moreover, be pointed out that the usual purpose of town and traffic planning is not merely noise abatement but rather the general improvement of environmental conditions (noise, safety, pollution, visual intrusion) and traffic (time and length of journeys).

It is with these constraints in mind that opportunities for action in the matter or urban and traffic planning will be discussed.

Urban Organisation

In the absence of any more refined analysis, the goal of noise abatement is taken to be the uniform reduction of average and peak noise levels, particularly in densely settled areas, while providing a contrasting choice of ambient noise quality marked by busy or quiet zones and periods. In mathematical terms, the objectives might thus be said to consist in (*a*) primarily minimising the sums of the products of population densities and corresponding traffic densities, sector by sector; (*b*) accessorily maintaining fairly substantial differences between a few partial products.

A definition of urban structures somewhat depends on the value which the town planner implicitly attaches to either partial objective. A fairly frequent aim in new residential neighbourhoods is to endow the centres with vitality through the automobile. Although other solutions than the motor car can be found, a minimum amount of background noise is an advantage which must be admitted, even in pedestrian areas. Extreme cases coming under the first partial objective are population densities so heavily concentrated that travel on foot need only be taken into account, or else semi-rural types of dispersion where costly road networks are required.

On a major scale, the density with which journeys are generated (or jobs, depending on the area) is found to be in ratio to population density, as has been observed in a large number of built-up areas, but length of journey is not allowed for and purely residential sections are solely affected.

In these sectors low residential densities are to be preferred in order to achieve a smaller product of population and traffic density, unless special precautions which no longer have anything to do with general planning are taken. A forecasting model of noise levels in terms of population density has been built in the USA.[22] A tenfold increase in density would cause ambient noise levels to increase by 12 dBA. In the case of the new towns in the UK, in practice the tendency has been towards small but moderately increasing densities.

As a general rule the urban planner combines such quite different objectives as less visual intrusion, pedestrian insecurity, pollution, noise and journey times into a single objective of environmental improvement. On the scale of the town this is a satisfactory approach provided the partial aim of reducing journey time is replaced by that of reducing individual journey lengths, since if the time of travel is reduced by increasing the speed, the result will be bad from a noise standpoint, unless most of the journeys take place on specially designed roads.

These considerations will contribute to a better understanding of the problems which arise according to the type of town organisation.

The Separation of Functions: The Linear Town

As far back as 1929, the Athens charter requested town planners to create sharp divisions between business, residential, recreational and traffic areas. A primary reason was the especially harmful nature of industrial establishments at the period. The 'linear' town, laid out along main arteries —with dwelling areas on one side and working areas on the other— adequately meets the objective of generally reduced noise levels: the corridor in between need only be carefully insulated from built-up and recreational zones. It is, however, difficult to organise centres of activity, the long main arteries are costly and the town is cut in two.

Among new towns developed according to the linear pattern may be mentioned Brasilia and Cumbernauld.

Radially Structured Towns

Linear towns offer greater development possibilities than radially structured towns, although most contemporary layouts are of this latter type. Development around a commercial or political central core results in the proliferation, beyond an initial ring composed of dwellings and local stores, of additional rings of a mixed industrial and residential nature. Owing to the absence of circular roads structuring the rings, none of the advantages of linear towns are present.

Distances to the centre are long, heavy lorries interfere with traffic on the periphery, and traffic lights must be installed on inadequate roads passing through residential areas. Insufficiently remote industries served by the road network further add to noise pollution.

Redevelopment attempted in these cases is such as to reduce noise disamenities; railway facilities serving the centre and a ring motorway restructuring the outer belts can ultimately endow it with the virtues of linear towns (such as the 'raccordo annulare' in Rome, which in relation to the city's size is far enough away from the centre).

The radial pattern is fairly acceptable in large metropolitan areas with adequate passenger transport facilities, particularly railways leading to the central core (New York, London, Paris, Hamburg, Copenhagen, etc.). While in medium-sized towns (e.g. of less than 600 000 population) the problem becomes much more acute, these are often equipped with express ring roads able to relieve the town of much of its inner traffic. This is true for most towns whose traffic has been reorganised (examples are the three Swedish towns of Västeras, Göteborg and Lund where, by prohibiting traffic through the centre, accidents, pollution and noise have been reduced).

Urban Grid Patterns

A road network designed as an unevenly meshed grid enables a town to go on developing along flexible lines and traffic to be efficiently 'distributed', which perhaps accounts for the grid's popularity in new towns. However, by thus promoting a diffuse traffic pattern, the danger is that noise levels may be unnecessarily increased, while little contrast is afforded between quiet and busy areas. Precautions which must be taken include a careful road hierarchy, facilities at intersections for slowing or restraining traffic on secondary roads, and zig-zag courses for such roads. When one-way roads are used journey length is proportionally increased, which may add to the general noise level.

Among towns which have adopted the grid pattern may be mentioned such various examples as Milton Keynes (1 km squares between motorways), Le Vaudreuil, France (300 m and 400 m squares), and Chandigarh, Pakistan. In the final account, grid patterns call for special care where road engineering and noise protection along the roadside are concerned.

In opposition to such grid patterns certain irregular arrangements provide interesting contrasts between sectors; the use of cul-de-sac networks thus enables a quiet environment to be achieved (e.g. the new town of Columbia).

Industrial Areas

Reserved industrial areas remain necessary, even in the case of largely non-polluting modern industries, which are still sources of annoyance owing to the motor vehicle traffic they generate (exceptions are relatively few).

Massive industrial areas, which earlier accounted for most of working life in small and medium-sized towns, seem now to be gradually disappearing; such areas generate journeys over fairly long distances and some degree of annoyance, in particular due to two-wheeled vehicle traffic in the early morning hours. Access routes to industrial areas which cannot be shifted should be appropriately redesigned, owing to the nuisance caused by both industrial and workers' vehicles.

With industries giving little pollution the industrial areas may be broken up. By thus scattering employment, journey lengths, and hence the time during which noise is generated, can be reduced. Relocation should preferably take place along main arteries affording suitable, non-polluting linkages with industrial enterprises. Examples are Runcorn and Redwood, which in this respect differ little from linear towns.

Open Spaces

Recreational and open spaces are located, according to case, in the centre of town and/or near residential zones. In the first case a sufficiently quiet environment is difficult to achieve unless a considerable surface is available. At least 10 ha are usually needed to achieve an average noise level under 50 dBA. Green strips along streams or around lakes unfortunately are often noisy even though the present trend is to provide open spaces encircling water surfaces, while it is difficult to take best advantage of pleasant sounds such as those made by fountains and waterfalls.

Interesting examples of open spaces large enough to promise some amount of quiet are found at Cergy-Pontoise, Isle d'Abeau, Brasilia and Vällingby, but in these cases access would have to be limited to quiet vehicles.

Hierarchisation of Urban Centres and Road Freight Terminals

Theoretically the number of vehicle-kilometres in a town can be minimised by increasing the number of centres of activity (as opposed to the single core found in traditional towns). In this connection two principles can be stated[23]

1. A commerical or service enterprise serving a population evenly distributed over a circular or rectangular area will generate a minimum amount of traffic if it is at the centre of the area.
2. In the case of enterprises of varying importance, travel can be minimised by placing them, for example, in centres of four types: city centres, sector centres (business and financial headquarters), suburban centres (supermarket facilities, etc., for populations between 50 000 and 200 000), and neighbourhood centres.

In the case of goods transport these principles must be slightly altered to take account of haulage and delivery costs:[24]

(a) An ideal town layout would be served by an inner ring road and include a common nucleus for goods distribution located inside the ring road. This nucleus should be located away from the centre at a distance amounting to some two-thirds or three-fourths of the town's radius;
(b) A large nucleus serving the entire town would not necessarily make general delivery costs any more expensive than a large number of geographically specialised nuclei.

For purposes of noise abatement these principles must be analysed in terms of overall journey lengths and types of road traversed.

The Tri-State Transportation Commission has studied goods distribution, in particular for Manhattan, and has concluded that 15% of trips by goods vehicles could be eliminated by suitably located freight terminals.

The fact is that noise would be far better controlled than by merely reducing journey lengths, since deliveries could be made by less polluting vehicles than the heavy vehicles used hitherto. One of the Tri-State Transportation Commission's comments was that much of haulage is performed by vehicles with a capacity of 8–10 tons which only deliver between 1 and 2 tons each day,[23] whereas under present conditions heavy vehicles emit some 10 dB more than light trucks.

Terminal organisation may impose certain constraints which private hauliers are unwilling to accept. The outlook is more promising in the case of large road or rail transport companies.

The location of road freight terminals on the town's outskirts is warranted if there is no acceptable access route towards the centre nor any suitable terminal site.[25] Rail transport offers a definite advantage in older cities possessing this facility.

Whenever the city has a ring road close to the centre it may also be preferable to locate road freight terminals near the ring road in an industrial area rather than any central location.

Urban Motorways

Public transport with its own right of way creates much less of a nuisance than private cars, provided quiet vehicles are used. Unfortunately no economic advantage is obtained unless passenger flows are considerable, at least in the present state of the art and economic context. Effectiveness from a noise standpoint is greatest when high-density areas are linked, which need not internally be served by private car but where walking or short-distance transport facilities suffice. Examples are urban centres of large metropolitan areas and more recently such new towns as Vällingby, which is connected to Stockholm by suburban railway. Usually, however, the tendency is to develop urban motorways rather than right-of-way transport facilities.

Urban motorways create considerable annoyance unless built on undeveloped land; otherwise people are induced in the medium and long term to move farther away from their place of work with the result that louder noise levels are increasingly diffused over ever-spreading areas.

It has been noted that for a given town the length of individual car journeys increases by $V^{0.75}$, where V is the average velocity in miles per hour over the road network. In towns developed over the long term, journey lengths increase at an even higher rate, the ratio being $V^{1.5}$. Taking the acoustic energy emitted by a vehicle as approximately proportional to the square of its speed, this means that the average increase in speed over an urban network results in considerably increased noise annoyance. If this is defined as the product of numbers of individuals and the amounts of acoustic energy they receive, the annoyance increases in terms of average speed over the network carried to the fourth power. This line of reasoning of course does not apply if the increase in speed primarily takes place along a few urban motorways with no dwellings located nearby. Generally speaking, however, the speed characteristics of urban motorways are often far greater than would be desirable.[26] To build a road allowing a speed to 110 km per hour instead of 50, results in a mere saving of 4 to 5 min over a distance of 7 km, whereas the noise increases by more than 10 dBA.

Among average-sized towns compelled to use urban motorways in addition to public systems, even though of an excellent kind, one which may be mentioned is Zürich.[27] Owing to traffic requirements it is planned to extend three expressways to meet in the city centre, forming a Y; 53% of the layout would be in cuttings and a considerable portion in tunnels or covered over. Environmental needs will call for additional work, putting the total cost of the project up from Sw.Fr. 936 million to 1102 million, or by 15%. By building car parks affording access to the centre on foot no interchange will be necessary, thus avoiding a considerable disamenity as well as injection into the centre of an unacceptable amount of traffic.

For thorough protection against noise from motorways these must be covered by or integrated with the urban structure, which can only be done in centres of exceptional density; medium densities are least favourable in this respect.

Pedestrian Areas and Cycle Lanes

Pedestrian areas created in new or old towns do a little to reduce average loudness, which reaches from 60 to 70 dBA in resonant streets and squares with high buildings and hard ground, where the sound of footfalls and some motor traffic is strongly reverberated. But since the noise is no longer of 'motor' but instead of 'human' origin, as a rule it is more easily accepted and even desired.

The servicing requirements of residents in pedestrian areas unfortunately lead to the limiting of vehicle delivery and service schedules to the morning or evening (as between 4 and 11 a.m. in the Strøget sector of Copenhagen). This can only be avoided by limiting the period of service-vehicle operation or by permitting a minimum amount of bus and delivery-vehicle traffic (as along Oxford Street in London); in this latter instance, however, noise is apt to increase since only noisy vehicles then circulate.

Public transport facilities are sometimes needed in pedestrian areas, either because there is no other alternative, as in the case of tram lines, or else because long pedestrian ways must be serviced.

Vehicles used in these areas, particularly if no others were allowed, would cause very little annoyance if electric or gas motors were adopted (like the 'Steyrbus' gas minibus planned for Vienna). Other systems, much like those found in fairgrounds or at airports, could also be used (business district of Miami, etc.).

Cycle lanes, which in European countries had largely declined, have become increasingly popular, especially in Great Britain, the Netherlands and Denmark, and should develop in line with the improved maintenance of existing cycle paths during coming years in all European countries. At Stevenage (UK), in 1966, 15% of all travel between home and workplace was by bicycle, thanks to a system of cycle paths sharply separated from the road network. The development of such routes, whether or not for the combined use of pedestrians, bicycles and motorcycles, may call for fairly wide lanes and thus later be made accessible for electric or invalid vehicles.[28]

Pedestrian facilities, which are sometimes neglected altogether and usually provided by pavements along carriageways, can of course be so dissociated from them as to afford pedestrians the advantage of safety and quiet; this, in fact, is what has been done in most new Scandinavian and French towns.

Improving a Particular Urban Sector

Improvements to urban sectors generally take place in the following stages:

1. The greater part of motor-vehicle traffic is diverted on to special highways.
2. Advantage is taken of improved traffic conditions thus achieved within the area by reorganising public transport, particularly bus traffic.
3. Car parks are built near the special highways.
4. General traffic is then restricted within the area by reserving streets for use by authorised vehicles, buses or pedestrians and by forming culs-de-sac.

If the noise level of buses is acceptable, ambient noise is thus significantly reduced. We have some interesting examples of improvement in urban sectors in Sweden[29] and in Great Britain.[30]

Reorganising some individual sector may, however, sometimes be hindered by the saturation of roads reserved for through traffic and by the need to improve the town's public transport facilities as a whole. On this account it may often be difficult to avoid restructuring an extremely wide area.

Specific Traffic Measures

Improving a town sector essentially depends on developing public transport facilities. If bus routes are protected through the use of bus lanes

TABLE 5.6
NOISE EFFECTS OF VARIOUS TYPES OF TRAFFIC CONTROL*

	Few heavy vehicles		Over 20% of heavy vehicles	
	Peaks	Average	Peaks	Average
Motorways and expressways				
Metering and diversion of traffic ⎫ Driver guidance ⎭	+−	+	−	+
Contraflow	−		−	
Provision of interchanges/dense traffic	−	−	−	−
Elimination of intersections/light traffic	+−	+−	+	++
Bus lanes	−	0	0	0
Noise barriers	+++	++	++	+
Urban arteries				
Widening of carriageways		+−	−	+−
Pedestrian lights			−	−
No stops at intersections	+−	+−	+	++
No parking		+−	−	+−
Green waves (progressive light timing)	+−	+−	+	++
Bus-actuated lights ⎫ Bus lanes ⎭	0−	0+	0	0
One-way traffic	−	−	−	−
No entry for heavy vehicles	+	0	++	++
Speed limits	++	+	+−	+−
Roads in residential areas				
One-way traffic	−	−	−	−
No entry for heavy vehicles	+	+		
Facilities for limiting speed and acceleration	++	+		
Bus services	−			
Cycle lanes	+	+		

* + = favourable effect; − = unfavourable effect; 0 = neutral effect.

and bus-actuated signals, average speeds can be approximately doubled. While noise levels may slightly increase, by controlling traffic lights the amount of stopping and starting and hence the noise peaks due to braking and acceleration can be reduced.

In dealing with traffic in general, the many specific measures which may be taken for its improvement do not invariably result in less noise. In a fair number of instances the increasing number of cars, even if traffic is not faster, causes noise levels to rise. Table 5.6 lists the measures most frequently taken and shows to what extent they may be expected to abate noise in the vicinity of roads. It does not therefore prejudge possible improvements for a group of sectors.

TRAFFIC PLANNING

Evaluation of a Transportation Plan
Normally, the engineer in charge of traffic organisation in an existing built-up area takes special care to reduce the duration of the drivers' trips. He sometimes assumes that his action will automatically lead to an improved environment and, as we saw, he has few evaluation factors where the noise criterion is concerned. Unfortunately his intuitions may be less fruitful than the town planner's: he may introduce longer routes for higher speeds and thus increase at once both the number of people involved and the noise levels. His main concern is with traffic at peak hours, whereas the acoustician is more preoccupied by evening and night traffic. We shall therefore examine the incidence on noise nuisance of the methods used to improve urban traffic.

Traffic Limitation and Speed Improvement
Quite a few local authorities thought that traffic limitations would reduce pollution, traffic jams, and also noise. The various types of limitation are unfortunately less efficient for the noise. For the same speed of the stream, eliminating one car out of two cannot reduce L_{10} or Leq by more than 3 dB and changes nothing in L_1, or LNP.

In fact, in congested areas a partial traffic limitation allows higher speeds thus increasing Leq and especially LNP and the dispersion of the levels. The level of the maximum L_{10} or Leq emitted by a road is not reached at the maximum traffic volume but at roughly half this volume. A significant improvement requires almost a total traffic prohibition or a rigorous speed limitation. Diverting the traffic to prescribed routes offers definite advantages, as pointed out on pp. 160–163. Aiming at maximum vehicle flow in an urban thoroughfare in the daytime is often a better objective than partially reducing the traffic. The cases when limiting the entry of vehicles might offer some advantages to reduce the noise, apply mainly to light traffic or heavy vehicles. In the first case L_{10} and Leq can be reduced by $10 \log Q/Q_0$, provided stream limitations are well observed (Q_0 being the traffic volume before, and Q after, the limitation). Eliminating trucks is always efficient for L_{10}, Leq, L_1 and LNP. We saw that a truck emits as much noise as 5 or 10 cars. Thus as far as noise is concerned, any action is more effective outside of the busy hours—which are not particularly annoying (pp. 98–100).

TABLE 5.7

THREE STREETS WITH SAME TRAFFIC VOLUME BEFORE AND AFTER A
NEW TRANSPORT PLAN

		L_{10}	Leq	σ	LNP
1	Before traffic jam	78·1	75	3·5	83·5
	After one way	80·8	74·8	5·6	88·8
2	Before traffic jam	81·5	76·8	4·4	87·8
	After one way	80	73·9	5·3	86·9
3	Before traffic jam	77·7	73·6	3·6	82·6
	After one way	80·2	74·4	5·2	87·4

When compared with an existing traffic block, any improvement allowing greater vehicle speeds would produce hardly any change in the noise level, as has been ascertained in Lyon. Only LNP increases significantly (Table 5.7). This can be explained as follows: for successive low accelerations in 1st and 2nd gears, a somewhat higher speed in 2nd and 3rd gear is substituted, so that on the whole the noise levels in Leq are hardly different (see Table 5.8).

TABLE 5.8

NOISE EMITTED BY AN R.16, in dBA (MOST APPROPRIATE GEAR RATIO)

Acceleration (m/s/s)	$\gamma = 0$	$\gamma = 1\ m/s^2$	$\gamma = 2\ m/s^2$
Speed km/h			
20	62	64	68
40	66	68	71
60	66–68	67–70	72

An experiment was carried out in Marseilles lasting 20 days, in connection with the creation of a north–south axis for public transport. Prohibiting car parking along the axis led to an increase of 33% in the average speed with the same traffic volume during the busy hours. While an important decrease in the carbon monoxide content of the atmosphere has been observed (a sign of the elimination of the traffic jam), the average level of the noise has not perceptibly changed.

Elimination of Crossroads—Regularisation of the Traffic

Let us recall that in a one-way street noise levels are higher downstream for vehicles accelerating after a crossroads. It is the heading car which starts with the highest and then noisiest acceleration. The inhabitants upstream are therefore comparatively less bothered. The length of the excessive noise zone depends on the cruising speed reached and on the nature of the vehicle (car or truck) (see Fig. 5.16).[31]

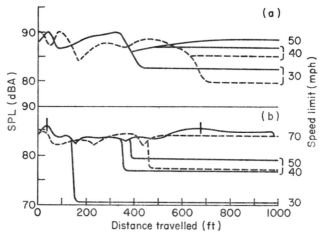

FIG. 5.16.   Effect of speed limits on noise of accelerating vehicles: (a) lorry and (b) car.
——— unladen, – – – fully laden (from Raff and Perry[31]).

Eliminating crossroad stops by means of underpassing or greenwave signalling may replace the noise due to the starting acceleration by the noise due to a gradient or a higher cruising speed which is felt both up- and downstream.

In the case of very light traffic, for instance at night, eliminating the starts may have a favourable effect provided this does not lead to speeds exceeding 60 km/h. In these circumstances every car is the heading car and starts with noisy acceleration. We may note also that the noise at a junction depends on each of the roads involved and eliminating the stops on only one of them may yield little improvement.

If crossroads are retained, limiting the starting acceleration and then the cruising speeds has beneficial effects (Fig. 5.17). It is difficult to plan for effective education of the drivers but planning road layout and design to limit the acceleration of the heading vehicle is possible. The traffic volume through a crossroad is actually the same if the heading vehicle accelerates at 1 m/s² or 2 m/s² but L_1 will be 5 dB higher in the second case. Layout and design for secondary roads may use large studs (as in Mexico City) or sharp bends in the road after the crossing.

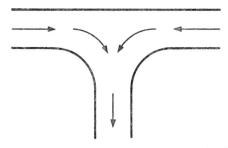

FIG. 5.17.   Sharp bends in the road limit acceleration.

As for the duration of the traffic light cycles, long cycles with minimum total red phases are favourable for lowering LNP. But it must be emphasised that the annoyance due to noises pulsed by signals cannot be adequately represented by statistical distributions (L_{10}, Leq, L_1, LNP, etc.) and it seems that certain time cycles are particularly unfavourable.[32]

One-way Streets

Resort to one-way streets hardly reduces the noise level, except for the inhabitant in the section before a crossroad. Unfortunately air pollution is higher in the same area; thus, in Lyon, waiting vehicles are kept on bridges before the crossroads. With one-way streets one or other of the following drawbacks are usually encountered:

1. An increase of speed and a consequent increase in Leq, L_{10} and LNP (Table 5.9).[33]
2. Longer routes, thus increasing the total acoustic energy emitted. Hence, the use of a one-way street in the example from Fig. 5.18[23] shows an increase in traffic volume from 900 to 1350 veh/h. Since the

roads are not saturated, average speed does not fall and the increase of Leq is about 2 dB. If the use of one-way streets sometimes relieves the pressure on certain roads, it is nevertheless against the principle of concentrating the traffic.

3. The diversion of traffic into previously quiet roads—which is particularly annoying when it includes heavy vehicles.
4. When buses are allowed to run on one-way streets both their stopping places and their routes are dispersed which causes a greater disturbance.

TABLE 5.9

	Streets	L_{10}	Leq	σ	LNP
Before	1	78	73·6	4·2	84·1
(two ways)		87	81·2	4·5	92·2
After	2	78·5	74·2	3·4	82·6
(one way)		85	79·5	4·0	89·5
	3	74·7	69·3	5·5	83·3
		81·3	76·2	4·7	88·2

Heavy Vehicle Traffic

Trucks or buses with diesel engines emit a high level peak noise, but its increase with the speed and acceleration is smaller than for cars, particularly for speeds under 60 km/h. Their contribution to the Leq level is important, all the more so as they are comparatively slow moving (for a truck: $\text{Leq}_H = L_0 + K \log S$, where $10 < K < 20$).

For heavy traffic with more than 10% trucks, eliminating crossroad halts may reduce LNP and Leq simultaneously up and downstream: (*a*) upstream, the braking noise may be eliminated—unlike cars, each truck

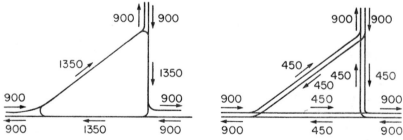

FIG. 5.18. One-way gyratory scheme, showing effect on traffic volumes (from Thomson[23]).

is very noisy in starting away, not merely the leading one; (*b*) downstream, a greater and constant speed may reduce Leq and LNP to some extent, but unless speed is limited it will have no effect on L_1.

Speed Limits

If the dispersion of the speeds is small and if there are no trucks, the noise reductions due to speed limits can be calculated as shown on pp. 95–98. In the other cases, the evaluation of the effect of speed limits requires the use of more accurate formulae. The noise level of a vehicle

TABLE 5.10
L_1 AND Leq VARIATIONS FOR DIFFERENT MEAN CAR SPEEDS

	Cars without trucks		Trucks only	
Mean speeds (km/h)	120–90	90–60	120–90	90–60
L_1	−5	−7	0	0
Leq	−4	−5	−1	−1

flow depends on the mean speed of the cars and of the lorries, as well as on the dispersion of their speed. Ullrich established the theoretical formulae for Leq and mean peak levels using the following assumptions:[34,35] a normal statistical distribution of speeds; tenfold higher noise emission for a truck than for a car at the same constant speed; an increase of the maximum noise level emitted by a car as 40 log *S*, by a truck as 20 log *S*.

Table 5.10 gives the reductions of L_1 and Leq for varying mean speeds of the vehicles and Table 5.11 for different speed limits. The initial situation

TABLE 5.11
Leq VARIATIONS FOR DIFFERENT MEAN CAR SPEED LIMITS

Speed limit	Cars only	10% Trucks	30% Trucks	100% Trucks
90 km/h for cars 80 km/h for trucks	−4	−2·5	−1·3	0
90 km/h for cars 60 km/h for trucks	−4	−4	−3	−2·4
60 km/h for cars and trucks	−9	−7	−5	−4

is assumed to be free-flowing car traffic with a mean speed of 120 km/h and a speed limit of 80 km/h for trucks.

Speed limits are very effective where there are no trucks. But for a stream of vehicles which includes more than 10% of trucks, moderate speed limitation will have no effect if it is not discriminatory, and the speed limits for trucks must be particularly low.

Trucks

Trucks have a particular responsibility for the increase in noise nuisance over a large part of the 24 hours and over a wide area. Heavy vehicle traffic on major roads at night is in some cases very dense and the quiet period may last only about four hours, or even as little as one hour. Such traffic is growing in outlying suburbs, increasingly equipped with traffic lights. The best solution to the problem would seem to be concentration and restriction of heavy vehicle traffic to a very few major roads.

Nevertheless, prohibiting daytime delivery traffic as a means of improving the normal traffic noise situation may be of questionable value. When such a prohibition is applied, for example to the milk delivery vans as in Paris, the solution consists in employing the British type of electric float. A similar problem arises with dust carts or with general delivery vehicles making frequent stops.

Conclusion

The potential of traffic engineering as a means of reducing noise is not negligible but does not unfortunately always coincide with transport requirements, nor is its efficiency easy to predict. Traffic diversion or severe speed limitation are the only means of reducing noise levels by more than 5 dBA.

URBAN ARCHITECTURE

Architects have a fundamental part to play to ensure quiet residential areas. It is their duty to make use of every simple consideration that relates reflection and diffraction phenomena to those governing geometrical optics. In complex cases, scale models or computer programs may be helpful, although they are likely to be rather costly.

Layout Plans and Housing Estates

Elementary solutions consist in transferring car traffic, and if possible parking, outside groups of buildings. These then surround quiet spaces

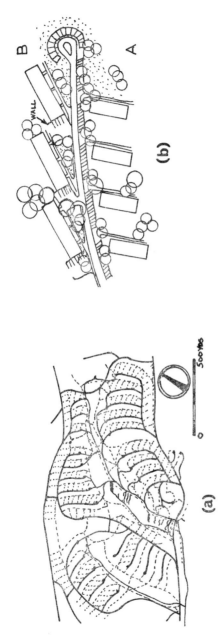

FIG. 5.19. (*a*) Clustered housing schemes, by Rüeichow/Eggeling, for Leverkusen-Steinbehel. (*b*) Loops permit quiet turning without reversing. Placing rows of dwellings with the ends to the roads reduces noise in front rooms. B disposition with a wall is a better one than A (from the *Architects' Journal*, Information Library, 13 February 1963).

and play areas on which the bedrooms look out. Such arrangements may entail the sacrifice of expected privileged orientations in relation to the view or the sun, although this will only occur on a few sites. Clustered housing schemes are preferred by many planners but they require longer networks or roadways, pipes, drains, etc. Nevertheless they eliminate cross traffic (Fig. 5.19). Their use is quite widespread in new housing estates of high standard or where the small size of the lots obviates the need for excessive increases in the lengths of the various networks.

Dispersed isolated buildings bring about a general diffusion of the traffic noise whereas buildings of former times which bordered the roads tended to limit it. A trend from 'open development' back to 'close development' may therefore be encouraged in high noise areas.

Buildings Close to Roads

Flats with a double exposure forming part of a building at right-angles to the road have no quiet room. In addition, the variation in the noise levels at the facades is much more sudden when the vehicles pass to or from the concealed portion of the road. If we take as a reference level the annoyance expressed by the inhabitants, it seems that siting a building perpendicular to the road results in an increase of 3 dB or so in the noise level.[38,39] From a purely physical point of view it needs to be stressed that facades perpendicular to the road do not help to reduce the sound transmission through the windows which in fact attenuate the sound waves more efficiently when they receive them at normal incidence.

These considerations tend to show that it is better to plan flats with a double exposure in buildings parallel to the roads, for then at least some of the rooms are quiet. In some cases, a slight projection of the gables may ensure one quiet facade (Fig. 5.19(b), p. 173).

Setting a building back from the road produces a comparatively feeble attenuation, particularly for the upper stories. To obtain an Leq level of 60 dBA at the facades of tall houses, they must be set back some 200 m, assuming a traffic flow of about 2000 veh/hour with 20% of trucks. Nomographs from Rathé set out the necessary distances to achieve different noise indices.[36] The traffic is assumed not to include trucks and the sound is propagated over grassland (Fig. 5.20).

The efficiency of screening buildings will generally be preferred to setting back. Immediately behind screened buildings other buildings of the same height can be erected. Table 5.12, extracted from a Swedish guide,[37] gives the necessary distances from the screen.

Two solutions can be considered. First, constructing high and long buildings parallel to the road creates quiet areas at the rear where houses can enjoy low noise levels even on higher floors (15 or 20 dB less than those observed without barriers). If it is not possible to erect insulated blocks of

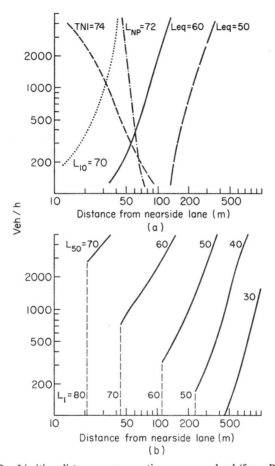

FIG. 5.20.   Limiting distances; propagation over grassland (from Rathé[36]).

offices, industrial buildings or high garages, well-insulated or properly equipped dwellings (e.g. with well-treated galleries) can be designed.

Secondly, low buildings may be used as barriers, so long as the acoustic protection thus achieved lies within the shadow zone. Such arrangements are particularly useful employing garages as barriers but they are also acceptable for a row of semi-detached dwellings which

Road Traffic Noise

TABLE 5.12

NECESSARY DISTANCES FROM THE BARRIER

Speed and type of screens	Height of buildings	Number of vehicles per mean annual 24-hour		
		2 500	10 000	40 000
90 km/h; screening building 3 m high, 40 m from middle of road	1 storey	at barrier	110	280
	3 storeys	80	150	300
	6 storeys	100	200	300
90 km/h; screening 12 m high, 40 m from middle of road	1 storey	at barrier	at barrier	at barrier
	3 storeys	at barrier	at barrier	at barrier
	6 storeys	60	60	60
50 km/h; screening building 3 m high, 40 m from middle of road	1 storey	at barrier	at barrier	110
	3 storeys	at barrier	70	150
	6 storeys	at barrier	80	200
50 km/h; screening building 12 m high, 40 m from middle of road	1 storey	at barrier	at barrier	at barrier
	3 storeys	at barrier	at barrier	at barrier
	6 storeys	at barrier	at barrier	at barrier

may be partially protected by low walls or earth-banks. Naturally, this solution is hardly applicable when the road lies well above the surrounding terrain.

When there are low buildings to be sited along the road, the town planner may profit generally by the effect of a shrubbery or cultivated area. If Leq must be less than 60 dBA the road to building distance can in this case be reduced from 200 to 100 m for a traffic density of 2000 veh/h, where the road is level with the terrain or slightly depressed.

Using buildings as barriers does not necessarily mean that they should be parallel to the road. U-shaped blocks, backed by the road, offer marked advantages over those giving on the road (Fig. 5.21). On the one hand, the buildings enjoy a pleasant quiet area and one avoids the increasing annoyance generally observed in buildings perpendicular to the road. On the other hand, multiple reflections of the sound create a reverberant field with a notably higher noise level than in open-air areas (3–10 dB more).

Various cases may occur according to the shape and size of the building lots. Figure 5.21 illustrates one alternative. The architect has placed two buildings perpendicular to a road carrying heavy night-traffic (it was thought that a facade perpendicular to the road would receive less noise). No consideration of view or of orientation justified this location. In fact, the second layout helps to secure quiet inside the square formed by the two buildings; each flat with a double exposure has a quiet side. It should also be noticed that the plight of buildings is worse when situated on the north side of the road, for the sunny facades are the noisy

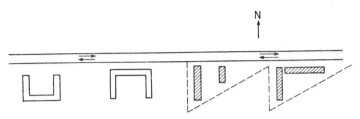

FIG. 5.21.   Layout of building near a road.

ones and acoustic insulation of the windows will render air-conditioning even more necessary. If it is impossible to form a continuous line of buildings along a road, the spaces between them need careful considera- tion. Lateral spaces should be less than the width of the buildings (Fig. 5.22). Side walls at an inclined angle or treated with absorbent materials may prevent the transmission of sound through multiple reflections.

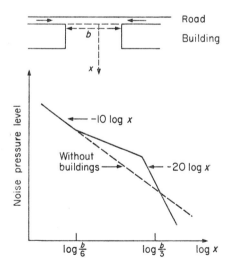

FIG. 5.22.   Schematic sound attenuation with discontinuous buildings (from Reinhold[14]).

Figure 5.23 shows the attenuation observed in the case of low-rise dwellings without a continuous line of buildings. A building at the rear is not necessarily helpful, for its facade may reflect sound.

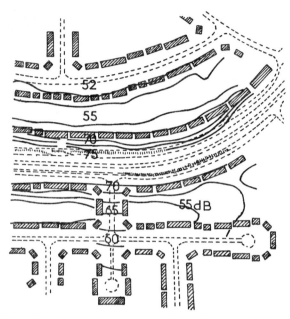

FIG. 5.23.   Plan of the site showing contours of L_{10} (18 h average) (from Watkins[40]).

Streets and Junctions

For an identical level of traffic, the reverberant noise level, Leq, shows a decrease of 3 dB when the street is twice as wide. This reduction is a little greater when the increased width allows room for plants or trees. The advantages of wide streets are still more obvious at junctions; oblique sides of buildings placed at the angle of the intersection also help to reduce the propagation of noise in the streets. As regards the height of the adjacent buildings, it only needs to be said that Leq and even L_1 are hardly affected by it, although of course when the upper storeys are set back they are quieter.

Elevated structures diffuse noise in the distance, particularly when the traffic passes near a waterway. Solid parapets may largely limit the distant propagation of the noise even if their height is limited to 1·5 m, although they may reinforce the noise level (by 3 dB or so) for houses flanking the road. However, this can be remedied by giving the parapet a suitable inclination.

Housing Design

Acoustic insulation of houses or schools bordering a street should only be resorted to in exceptional cases, for it does not ensure proper habitability outside. The absence of air-conditioning may nullify the effect of insulation if the facades are exposed to the sun. Insulation treatment can offer some advantages in very cold countries but it is costly and not very efficient for detached houses with gardens. Insulating windows may lead to a considerable reduction in the noise level; attenuation of 40–45 dB can be achieved with double windows and 25–30 dB with good-quality well-fitting windows equipped with thick glass panes (1 cm). Unfortunately, ventilating louvres reduce the insulation. In Britain, the Building Research Station has carried out a thorough investigation using a flat situated near the Midland links motorway.[41] Thermal comfort was ensured by means of specially developed ventilating devices and venetian blinds. With double

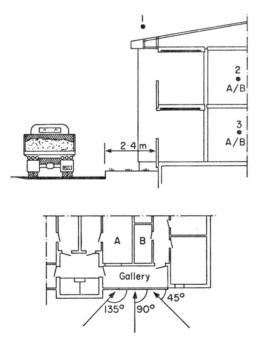

FIG. 5.24. A gallery for the protection of rooms (from Gustafsson and Einarsson[42]).

windows and double french doors it was possible to obtain a level of 45 dBA inside the flat, as against 80 dBA at the external facade. Acoustic absorbents placed on the sides and the ceilings of loggias reduce the noise of rooms set-back (Fig. 5.24). For residential buildings, outer or inner corridors along the facades are only suitable for flats with a single exposure. Failing this, the kitchen, bathroom, hall and lumber-room can be located along the corridors. A Swedish study compares the efficiency of various devices against noise from truck traffic. Figure 5.24 shows that the sound level in the rooms A and B may be reduced by 4·5 dBA (Table 5.13). A solid balustrade and efficient absorbents on the ceiling of the gallery offer obvious advantages.

TABLE 5.13

EFFICIENCY OF VARIOUS DEVICES

Device	dBA
1. No barrier; no absorbent treatment	0
2. Air-proof barrier with a height of 1·1 m	2·5
3. No barrier; absorbent of medium quality (50 mm wood wall board)	1·2
4. Air-proof barrier; absorbent of medium quality	3·2
5. No barrier; absorbent of high quality (50 mm wood wall + 50 mm mineral wool)	1·7
6. Air-proof barrier; absorbent of high quality	4·5

CONCLUSION

The means for improving our environment at the architects' disposal are among the most effective, whereas the perfection of vehicle technology or traffic engineering can yield only modest gains, say about 5 dBA less in the near future. As against this, town planning and careful orientation of housing may easily achieve a reduction of 15 dBA, and sometimes as much as 20 or 30 dBA in noise levels. Unfortunately, such arrangements may be expensive and can be applied only to new building. This is why everyone—road and traffic engineers, architects and town planners—should be engaged in the campaign against noise, and in particular be aware of the possibilities offered by their various techniques.

REFERENCES

1. *Environmental Implications of Options in Urban Mobility*, OECD, Paris, 1973.
2. Swetnam, G. F. and Murray, W. S. Feasibility study of noise control modifications for an urban transit bus, *Internoise*, 1973.
3. Schuber, H. Lärmminderung bei Betriebsmitteln und Anlagen des öffentlichen Personennahverkehrs, *Kampf dem Lärm*, No. 571, October 1971.
4. 'California Steam Bus Project', Final Report, January 1973.
5. Elektrostrassen Fahrzeugen, *Kampf dem Lärm*, No. 571, October 1971.
6. Jonasson, H. G. Diffraction by finite impedance wedges, *JSV*, 25 (4), 1972.
7. Schreiber, L. Stand der Bemühungen um einen ausreichenden Mindestschutz der Bevölkerung gegen Verkehrslärm, *Kampf dem Lärm*, 3 June 1973.
8. Ozanne, F. Le Boulevard Périphérique dans la Traversée du Secteur Zonier, No. 9, *Travaux*, March 1973.
9. Rucker A. and Gluck, K. Bauliche Schutzmassnahmen für Minderung des Strassenverkehrslärms, *Strassenbau und Strassenverkehrstechnik*, No. 47, 1966.
10. Fleischer, F. Zur Anwendung von Schallschirmen, *Lärmbekämpfung*, No. 6, December 1970.
11. Jonasson, H. G. Sound reduction by barriers on the ground, *JSV*, 22 (1), 1972.
12. Scholes, W. E., Salvidge, A. C. and Sargent, J. W. Field performance of a noise barrier, *Journal of Sound and Vibration*, 16 (4), 1971.
13. Reinhold, G. and Burger, W. Die funktionnelle und betriebliche Erprobung absorbierenden Lärmschutzwände an einer Autobahn, *Strasse und Autobahn*, No. 1, 1971.
14. (a) Reinhold, G. Bau und Verkehrstechnische Massnahmen zum Schutz gegen Strassenverkehrslärm, *Strassenbau und Strassenverkehrstechnik*, No. 119, 1971; (b) Rapin, J. M. Protection phonique aux abords de voies rapides urbaines. Mesures de l'influence du profil en travers et de la présence d'écrans et d'absorbants sur la pression acoustique aux abords des voies rapides urbaines, *Cahiers du CSTB*, No. 109, May 1970.
15. Dreyer, P. 'Barrières de Protection contre le Bruit de Trafic Terrestre', Rapport interne, IRT-CERN, July 1973.
16. (a) Kurze, U. J. Noise from complex road traffic, *JSV*, 19 (2), 1971; (b) Kurze, U. J. and Anderson, G. S. Sound attenuation by barriers, *Applied Acoustics N.L.*, 4 (1), 1971.
17. Green, W. R. 'A Cooperative Program to Control Highway Noise', Region IV AASHO Operating Subcommittee on Roadway Design, 19 July 1972.
18. Harmelink, M. D. and Hajek, J. J. 'Evaluation of Freeway Noise Barriers', Ontario Ministry of Transportation and Communication Highway Research Record No. 448, 1973.
19. Alexandre, A. 'Le Bruit des Véhicules à Moteur', OECD.
20. Reinhold, G. Die Wirkung von Abschirmeinrichtungen zur Lärmminderung an Strassen, *Strassenbau und Strassenverkehrstechnik*, BAS Köln, No. 157, 1974.
21. Hauskins, J. B., Jr. The kinematic sound screen: an effective solution to highway noise abatement, *Highway Research Record*, No. 448, 60, 1973.

22. *Transportation Noise Pollution. Control and Abatement.* NASA, Langley Research Center and Old Dominion University, USA, 1970.
23. Thomson, J. M. *Methods of Traffic Limitation in Urban Areas*, OECD, U/CKO/72.700, Paris, 1972.
24. 'Systèmes de Distribution Urbaine de Marchandises', Bulletin de l'I.R.T., No. 5, January 1974.
25. *Rail/Road Terminals. Road Freight Terminals.* Société de Recherche et d'Economie Appliquée, OECD, 1970.
26. Bor, W. and Roberts, J. Urban motorways impact, *The Town Planning Review*, No. 4, October 1972.
27. Buchli, G. 'Zurich and its Traffic Problems, with Specific Regard to the Planning of Expressways', Symposium on Transportation and Environment, University of Southampton, April 1973.
28. Perraton, J. K. Planning for the cyclist in urban areas, *The Town Planning Review*, **39** (2), July 1968.
29. 'Principles for Traffic Re-planning with Respect to Road Safety', Report 55, Institute for Urban Planning, Chalmers University of Technology, Göteborg, 1972.
30. Bor, W. Urban transport and environment, *JSV*, **15** (1), March 1971.
31. Raff, J. A. and Perry, R. D. H. Vehicle noise studies, *JSV*, 8 June 1973.
32. Lamure, C. 'Human Reactions to Traffic Noise', Symposium on Noise in Transportation, Southampton, July 1974.
33. Pachiaudi, G. *Bruit dû à la Circulation Urbaine et Eléments de Plan de Circulation*, IRT, December 1973.
34. Ullrich, S. Geschwindigkeitsbeschränkungen, ein Mittel zur Reduzierung des Lärmes von Autobahnen und Schnellstrassen?, *Strassenverkehrstechnik*, No. 1, 1973.
35. Ullrich, S. Der Einfluss von Fahrzeuggeschwindigkeit und Strassenbelag auf den energie äquivalenten Dauerschallpegel des Lärmes von Strassen, *Acustica*, No. 30, 1974.
36. Rathé, E. J. Review of transportation noise in residential areas, *Internoise*, 1973.
37. Urban Planning and Noise from Road Traffic (General guide to planners), National Swedish Board of Urban Planning.
38. Bacelon, M. and Lamure, C. La gêne dûe au bruit de la circulation automobile, *Cahiers du CSTB*, No. 762, October 1967.
39. Bietry, J, and Aubree, D. Le bruit des rues et la gêne exprimée par les riverains, *Cahiers du CSTB*, No. 1174, April 1973.
40. Watkins, L. H. 'Traffic Noise', Symposium on Transportation and Environment, University of Southampton, April 1973.
41. An experimental investigation of motorway noise and sound insulation alongside the Midland links motorway, *Architects Journal* (to be published).
42. Gustafsson, I. and Einarsson, S. Gallery houses with respect to traffic noise, *Internoise*, 1973.
43. 'A Design Guide for Engineers', Highway Research Board, Nat. Acad. Sci., Washington, D.C., 1971.

CHAPTER 6

REGULATORY AND ECONOMIC INSTRUMENTS FOR TRAFFIC NOISE CONTROL

A. ALEXANDRE and J.-Ph. BARDE

NOISE EMISSION LIMITS

At international level, the United Nations Economic Commission for Europe adopted in 1968 uniform provisions concerning the approval of vehicles with regard to noise. The measurement procedure is that developed by the International Standards Organisation (ISO Recommendation R362), which consists of measuring the maximum noise possible under normal town driving conditions, i.e. during acceleration at full throttle in an intermediate gear, starting from an engine speed corresponding approximately to the speed of maximum torque (the initial speed before acceleration being limited to 50 km/h).

The Council of the European Communities (Common Market), moreover, laid down a directive on 6 February, 1970 to harmonise the legislation of EEC member states regarding acceptable noise levels. So far this directive does not cover two-wheeled vehicles. The method of measurement, again, is that suggested by ISO Recommendation R362.

The acceptable maximum noise levels fixed by the Council of the European Communities are shown in Table 6.1. These limits are now compulsory in the nine Common Market countries.

But discussions are going on in the Working Party on Vehicle Construction of the Common Market, in order to adapt methods of noise measurement more closely to the different driving conditions encountered in urban areas and to the different types of vehicle (automatic transmission vehicles, for example). Certain countries would prefer a method in which noise is measured on stationary vehicles so as to speed up and simplify the periodic checks on vehicles. Some would like current regulations, which take account only of noise emitted during maximum acceleration, to be supplemented by measurements of noise emitted at cruising speed so that annoyance due to the noise of traffic on highways and urban motorways can be better allowed for. It might also prove interesting to develop a

TABLE 6.1

DIRECTIVE BY THE COUNCIL OF THE EUROPEAN COMMUNITIES (COMMON MARKET) RELATING TO MAXIMUM PERMISSIBLE NOISE LEVELS (NEW VEHICLES)

Class of vehicle	Acceptable noise levels* (dBA)
Passenger vehicles with seating capacity for not more than 9 persons including the driver	82
Passenger vehicles with seating capacity for more than 9 persons including the driver and a maximum permissible weight not exceeding 3·5 tons	84
Goods vehicles with a maximum permissible weight not exceeding 3·5 tons	84
Passenger vehicles with seating capacity for more than 9 persons including the driver, and a maximum permissible weight of more than 3·5 tons	89
Goods vehicles with a maximum permissible weight of more than 3·5 tons	89
Passenger vehicles with seating capacity for more than 9 persons including the driver and powered by an engine of 200 hp DIN or over	91
Goods vehicles powered by an engine of 200 hp DIN or over and having a maximum permissible weight of over 12 tons	91

* To allow for uncertainties in the measuring instruments, the result of each measurement made at a distance of 7·5 m shall be the instrument reading less 1 dBA.

standard cycle of noise emission by motor vehicles in urban areas, similar to that which has been done for air pollution (*see* p. 101). The practice of measuring in dBA could well permit an increase of noise in the low frequencies, which would lead to very serious problems for town planners and architects, low frequencies being extremely difficult to attenuate. Some specialists therefore suggest that sound pressure levels throughout the frequency spectrum should be covered.

As regards lowering noise limits in the future, it should be noted that long-term recommendations were submitted in the UK by a working party on road traffic noise, the aim being to lower noise standards gradually to 75 dBA for private cars and to 80 dBA for commercial vehicles.[1] This working group has, however, emphasised that it will be particularly difficult to attain such standards for commercial vehicles and that several years of research will be necessary before engines and vehicles are designed to meet such standards. The group therefore considers that the entire fleet of commercial vehicles is hardly likely to be replaced by one less noisy before 1985.

In the USA, it is worth noting that Chicago adopted an ordinance in 1971 concerning maximum permissible noise levels.[2] This ordinance prescribes such a programme for reducing vehicle noise, that by 1980 every new vehicle—whether motorcycle, heavy truck or private car—should emit no more than 75 dBA when measured at a distance of 15 m (equivalent to 80 dBA at 7·5 m, the measurement distance adopted in Europe). The details of these restrictions and the timetable appear in Table 6.2.

TABLE 6.2

ORDINANCE ADOPTED IN CHICAGO FOR GRADUALLY
TIGHTENING STANDARDS OF NOISE EMISSION

Date of construction	*Maximum limit (dBA)*
Motorcycles	
Before 1 January, 1970	92
After 1 January, 1970	88
After 1 January, 1973	86
After 1 January, 1975	84
After 1 January, 1980	75
Vehicles heavier than 8 000 *pounds*	
After 1 January, 1968	88
After 1 January, 1973	86
After 1 January, 1975	84
After 1 January, 1980	75
Private cars and other motor vehicles	
Before 1 January, 1973	86
After 1 January, 1973	84
After 1 January, 1975	80
After 1 January, 1980	75

According to the UK proposal and this ordinance regarding the reduction of noise limits:

1. Noise from the noisiest vehicles (motorcycles and heavy trucks) would be reduced more quickly and to a greater extent than that from private cars in order to eliminate peaks standing out from the background noise.
2. In the long term, noise from every vehicle would be reduced by at least 10 dB (amount varying according to category), so as to significantly reduce the average noise level from traffic.

THE UK LAND COMPENSATION ACT OF 1973*

Implications of the Act

The conflict between the community's requirement for new roads, airports, etc., and the private individual's right to enjoy his home and garden undisturbed is recognised as a very important problem. There is now a need for a better balance between the necessary provision for the community as a whole and the mitigation of harmful effects on the individual citizen. To redress the existing imbalance against the interests of the individual, new development should be planned so as to minimise the disturbance and destruction which would otherwise be caused, and the compensation code be improved in order to alleviate any remaining distress.

This new emphasis on better planning and design with the objective of minimising noise, fumes and other forms of pollution, together with enhanced compensation payments, is reflected in the UK Land Compensation Act (1973).[4] On the basis that 'the polluter pays' it is considered that these additional costs would have to be borne by the community through the authorities responsible for the development, such as highway authorities, airport operators, statutory undertakers and the government itself.

We are, however, only concerned with three aspects of an Act which deal also with other aspects of compensation. First, the powers of highway authorities to acquire land additional to that required for the scheme and to carry out certain works on that land to mitigate physical factors; second, the provision of grants to insulate properties affected by noise

* This section draws heavily on a paper presented to the OECD.[3]

from the construction of the public works or their use; third, the compensation which may be paid where the value of certain property has depreciated as a result of noise or other physical factors arising from the use of certain public works.

Land Acquisition

Highway authorities are empowered to acquire land other than that which is specifically required for the road scheme in question, and execute works on their land for mitigating any adverse effect which the road scheme—or its use—will have on the surrounding area. Under these powers, highway authorities can undertake works such as the erection of acoustic barriers or the construction of earth banks to minimise noise nuisance.

These provisions arise out of a recognition that *prevention is better either than cure or compensation*. It enables a highway authority to deal with the noise and other problems in the most economical way. And it is only when the mitigating works are not a complete solution to the problem that the need for sound insulation or compensation arises.

The cost of land in the UK is very high, and in view of the great pressures on land space there is a limit to the extent to which these powers can be effectively used. Nevertheless, by far the single most important element in the annual costs of implementing this Act is assumed to be the acquisition of land and associated mitigating works. This accounts for some £30 million a year (plus another £10 million a year for improved design), out of total expenditure of some £55 million in connection with the roads programme.

Sound Insulation

The sound insulation provisions are, of course, specifically directed at noise. But they are complementary to the compensation provisions because in assessing compensation it has to be assumed that betterment in the form of sound insulation or other noise-mitigating works has been installed where it will be made available (whether or not it is, in fact, installed).

The Act gives the Secretary of State general powers to make regulations imposing a duty or conferring a power on responsible authorities to insulate buildings against noise caused or expected to be caused by the construction or use of public works, or to make grants towards the cost of such insulation.

Effect has been given to this provision by the Noise Insulation Regulations 1973 which concentrate on the chief problem of traffic noise.[5] The regulations require highway authorities to carry out, or make grants for the cost of work for insulating certain residential properties. Insulation is mandatory where the following conditions are met: (a) a new, or substantially improved, road is constructed, and (b) the road is first opened to public use after 16 October 1972, and (c) properties are, or will within 15 years, be subjected to increased traffic noise of at least 1 dBA and to a total noise level of 68 dBA or above on the L_{10} (18 hour) index.

Rooms eligible for insulation are living rooms and bedrooms on any facade of the building which are exposed (at 1 m in front of the most exposed window) to noise at or above 68 dBA. Discretionary powers are given to provide sound insulation for properties similarly affected by noise from new or improved roads first opened to traffic after 16 October 1969.

The highway authority is also given discretionary powers to provide sound insulation for dwellings where substantial construction noise (normally in excess of 70 dBA) is otherwise likely to be experienced, even though the road when in use may not generate sufficient noise to justify the provision of insulation. Whether or not insulation is provided in these circumstances depends on a number of factors including the noise levels, the duration of the construction period and the hours during which work is carried out.

The specifications for insulation work are very detailed but basically the package consists of double glazing, supplementary ventilation and, where necessary, venetian blinds and double doors. The cost of the package depends upon the circumstances of the dwelling in which it is installed.

The qualifying level of 68 dBA was chosen after surveys in the UK had shown that 70 dBA on the L_{10} (18 hour) index seemed to represent the limit of what was tolerable for the majority of the population; it is not, however, regarded as an acceptable maximum limit (see Chapter 2). It was considered that 68 dBA provided sufficient tolerance on this to ensure that anyone suffering from noise at or above 70 dBA would obtain insulation. There are, of course, arguments for adopting a lower level, but such is the extent of the traffic noise problem in the UK that the costs of providing sound insulation for those affected by noise from new and substantially improved roads at or above a 65 dBA standard would be approximately double the present cost of £6 million.

There will inevitably be pressures to relax the existing qualifying level and to apply the principle more generously. There are already pressures to apply it to dwellings affected by noise arising from the intensification of use of existing roads. While more generous standards may be economically and socially acceptable at some future date, the costs of extending the principle to properties affected by noise from existing roads make it necessary to consider the implications very carefully. It has, for example, been estimated that the cost of *sound insulating all residential properties affected by noise of 68 dBA or above from existing roads would be of the order of* £1000 *million!*

Expenditure of this order on what is no more than a *palliative* would be a very incomplete answer to the problem of traffic noise. Despite the undoubted benefits of sound insulation in providing tolerable living conditions in many existing properties, the noise levels in certain rooms of insulated properties could still be undesirably high. Moreover, sound insulation only deals with the problem of noise inside the property. It can do nothing to mitigate the nuisance in the garden.

The real value of the sound insulation package has to be seen against the much wider background of the alternatives which might face the government if nothing were done.

Whatever pressures there may be for the wider provision of sound insulation, there is no doubt that those with the most compelling case for protection are those newly subjected to a significant noise source. Failure to respond to the needs of this group by better design treatment and, where this is not practicable or economic, by installing sound insulation, might lead to political protest and to demands for rehousing by those worst affected. In these circumstances, the government might well be faced with the prospect of paying compensation on a vast scale for thousands of useless houses and either converting them to some other use or demolishing them and clearing the sites. The problem would in many cases be exacerbated by the need to find new housing land. This is an almost insuperable problem in certain parts of the country. It would be wrong, however, to overstress these problems because it is by no means certain that the postulated political pressures would develop on a large scale, but the dangers would certainly exist once the dam was breached in one place. The conclusion might be that sound insulation, while undoubtedly expensive, is a relatively cheap and very worthwhile option for the country, while at the same time benefiting many people whose living conditions would otherwise be intolerable and allowing them to exercise the choice of remaining in their existing homes rather than being uprooted.

The principle of sound insulation is therefore considered to be one of limited application. The most cost-effective way of dealing with vehicle noise will continue to be to *find ways of reducing it at source and of segregating traffic and people.*

Compensation

Many public developments, such as new roads and airports, can cause depreciation in the value of property (described as 'injurious affection' in the Act); in some cases the property owner has no redress at law. In such cases the Act provides a new statutory right to compensation where depreciation exceeding £50 is caused by physical factors arising from the use of the public works. These physical factors are specified in the Act as 'noise, vibration, smell, fumes, smoke and artificial lighting and the discharge on to the land in respect of which the claim is made of any solid or liquid substance'. For the purpose of this section, however, the physical factor of importance is noise and the other factors are for convenience largely ignored. The compensation is payable mainly to owner-occupiers of properties (houses, small businesses, small farms—and to landlords of dwellings) where the interest in the property was acquired before the start of use of the public works. The amount payable is assessed on the depreciation in the market value arising from the use of the works. To eliminate the effects of temporary construction work and other short-term fluctuations the value is allowed to stabilise for 12 months after the start of use of the works. Previously, under the Compulsory Purchase Act 1965 (which remains in force), compensation could be paid only in limited circumstances where there was an interference with a legal right in land (such as a right of way or of light) caused by the *construction* as opposed to the *use* of public works constructed on land compulsorily acquired.

There is clearly much virtue in the principle of compensation based on the depreciation of a person's stake in his property, where that depreciation occurs as a result of physical factors. Although no compensation is payable for business losses, the compensation payments particularly to house owners do go a considerable way to remove what might otherwise have been an insuperable financial barrier to moving elsewhere. It therefore does much to restore to the house owner the freedom of choice to stay or go that he would have had before the noise source was inflicted on him. If he is unwilling to tolerate a noise or the other physical factors arising from the new works he is financially able to leave, although the compensation payable would not cover legal and other costs involved in moving.

Compensation will be assessed on behalf of the authorities involved either by the district valuers (who are employed by central government), by local authority valuers or by valuers appointed by the public authority carrying out the work. Which valuer is responsible for assessing the compensation depends on who is responsible for constructing the public works giving rise to compensation. Despite the undoubted expertise of the valuers, valuation is not an exact science and the value of property is often a matter of opinion rather than fact. Accordingly it is anticipated that disputes may well arise about the levels of compensation offered; this is especially likely in the early days of the Act before property owners and their advisers become familiar with the valuation methods and principles. Property owners will, of course, employ surveyors and valuers to formulate and negotiate claims on their behalf. Where compensation is payable professional fees may be reimbursed. The Act also makes provision for appeals to the Lands Tribunal, which is an independent judicial body especially set up to deal with such issues.

Some of the problems of valuation are undeniably difficult, though they are certainly not insuperable. One of the major problems will be that of ensuring equity of treatment. Each property is different and will be affected by noise and other physical factors in a different way. It is, therefore, necessary to consider the early cases very carefully indeed since these will inevitably tend to set a precedent for treatment of the cases which follow.

Compensation is payable only for physical factors arising from the source of the new works. Thus it is necessary to disregard such factors as visual intrusion and increased traffic flows on an unaltered existing road. If account had to be taken of this latter factor it would lead to many arguments about where eligibility for compensation begins and ends and who was eligible for it and who was not.

The Act does not require a before and after works valuation, but a valuation which assumes that the works are there and which assesses depreciation only for the effect of the physical factors resulting from the use of those works. In other words, if a motorway generated only silent traffic, visible but causing no vibration, smell, fumes or smoke, no compensation would be payable. The analogy is with the sort of actionable nuisances which could give rise to claims for damages if committed by the private sector. But visual intrusion is not actionable and is also not included in the list of physical factors for which compensation might be payable.

As can be imagined, the valuers will obtain guidance on current values from properties actually bought and sold in the relevant area. It is current values that are important because the Act requires that compensation shall be assessed by reference to prices current on the first day of the claim period, i.e. one year after the start of use of the works in question.

Compensation is limited to depreciation caused to the *existing use* value of the land, that is the value of a house as a house. There is no compensation for depreciation in its value for potential development, for example, for conversion to offices or for building another house on part of the garden. The underlying concept of Part I was to relieve hardship caused to people in actual occupation of their property and not to compensate them for reduction of some increased value the property might hypothetically have possessed if it could have been used for some different purpose. Broadly speaking, compensation is not payable in the UK for refusal of a planning permission or for conditions or restrictions placed upon it, and to have allowed claims for depreciation to hypothetical development values could have undermined this principle and encouraged unmeritorious claims based on arguments about the hopes of future planning permissions. On the other hand, public works may enhance as well as depress values and provision is made to set-off any increase in the value of the claimant's interest in the property itself or his other contiguous or adjacent land as a result of the works. For example, the value of a house, in spite of the effect upon it of the noise and other physical factors arising from the construction of a nearby motorway, might in fact be maintained or even increased because of improved accessibility. In such case no net depreciation has been caused and no compensation will be payable. In considering such benefit an increase in value for development is not ruled out but the authority would have to be able to establish that such enhanced value was really there and that it resulted from the public works to which the claim relates. This principle of set-off for betterment has been applied in the UK to compensation paid where land is compulsorily acquired for many years.

In conclusion, it is right to emphasise that the compensation provisions are regarded as the *final means of redress* when other positive measures (such as sound insulation, noise barriers, better planning, etc.) to minimise the adverse effects of the public works have all been taken.

REGULATORY INSTRUMENTS AND ECONOMIC INCENTIVES

A noise abatement policy implies the existence of a number of complementary actions interrelated in a complex system. Not only should the policy aim to reduce noise at source but it should also require action along transmission paths (since propagation depends on road design, urban planning, protective devices, etc.), at the reception point (the orientation and sound insulation of buildings), and even with respect to the transport medium itself (behaviour of car users and improvements in public transport).

Similarly, different agencies are involved, such as motor vehicle manufacturers, the users of vehicles and public authorities. The question is: which kinds of regulatory instruments can be used by public authorities so that they will have maximum impact on the various parameters and influence the different agencies concerned, so bringing about a satisfactory reduction of noise?

Traditionally, one instrument has been used: direct controls in the form of emission standards, building regulations, and so on. But in many instances such controls appear to be rather cumbersome and costly to implement. Economists have cast serious doubts on their actual efficacy as compared to an economic approach based on charges related to noise emission levels.

The assumptions underlying this economic approach are quite simple and have already been outlined in Chapter 3. Since pollution problems exist because there is no market for externalities, the solution could be the setting of a price on noise by public authorities and then leaving the economic system free to reach an equilibrium. In Fig. 6.1 the optimum

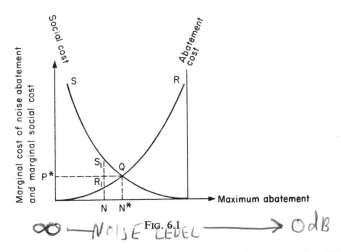

FIG. 6.1

price is achieved by requiring the emitter to pay a charge equal to P^* per decibel emitted. At a level below P^* it is cheaper for him to reduce the noise than to pay the charge, above P^* it is cheaper to pay the charge. The level of noise reduction is therefore fixed at N^*.

Thus, the car manufacturer and/or the car user could be subjected to a noise charge. The road developer could also be made liable for the full social cost he may impose. As we have seen, the Land Compensation Act

1973, recently adopted in the UK, requires the agent responsible for road building to give financial compensation in certain circumstances to those persons living alongside the road for the noise nuisance they may have to bear after the road is built.

The question of emission charges is therefore by no means a purely theoretical or academic one. Water pollution charges have been implemented in several countries, particularly in France, the Netherlands and Germany, and have been made to yield quite satisfactory results. On the other hand, the very notion of a noise emission charge encounters strong opposition and a number of counter-arguments. It is therefore worth giving some consideration to the various arguments for and against charges, in the light of their application to the traffic noise problem.

The Case Against Charges

The economic hard-liners who advocate 'zero growth' and other 'neo-Rousseauists' claim that the charge is only a way of paying for the right to pollute and degrade, i.e. 'I can pollute because I pay', which is an attitude not to be tolerated.

Hence, they say, the problems must be solved by absolute prohibitive measures alone, regardless of the cost. This argument can easily be refuted; if the rate of the charge is 'effective' enough as defined previously (level P^*), the best result will be achieved automatically, and at least cost to the community. Everything depends on the price: if it is too high it will be much like any ordinary fine, if too low it will be ineffective, since it will be more advantageous to pay than to stop the noise or pollution (it would certainly then be a case of buying cheap the right to pollute). For the sake of simplicity, the term 'to pollute' will be used to describe any act detrimental to the environment.

Among the more formidable opponents of economic mechanisms to combat pollution are certain government officials responsible for issuing and supervising the application of direct regulations concerning pollution. For example, regulations prohibiting plants from discharging more than a given volume of noxious materials, or motor vehicles from emitting noise above a given decibel level. This is known as the *standards or direct regulation approach* and its adherents—who may be called 'regulators'—put forward both 'good' and 'bad' reasons.

Prominent among the 'good' reasons is the argument that no optimum rate for the charge can be determined, since no one knows how to calculate the social cost or the disamenity: how can the cost of noise to the community be determined? Indeed, how can the cost of annoyance and insomnia be calculated? This first argument is a good one because the difficulties of calculating social cost are indeed enormous (see Chapter 3). But the answer is that we need not bother to determine the optimum rate. We merely calculate the rate not in terms of social cost, but in terms of the marginal cost of reducing noise.

The polluter will accordingly reduce the noise to the point where the rate of the charge is equivalent to the cost of reduction, i.e. the point where it is cheaper to reduce the noise than to pay the charge. We will return to this point later.

Another 'good' reason is that the charge leaves the results somewhat uncertain: unlike the standard, which fixes a compulsory ceiling, it is not known when or how the expected result will be achieved.

Let us now examine the 'bad' reasons. First, regulators assert that regulation is a classical legal and administrative system that has proved its worth; what should be done, they say, is to stick to familiar procedures. But in saying this, they fail to understand that the nature and dimensions of the problems have so changed that traditional regulatory methods may well prove extremely costly, cumbersome and ineffective.

Another 'bad' reason, in this case an implicit one, is that regulations provide government officials with powers that they have no desire whatever to relinquish: that of issuing regulations, of assessing effectiveness, of weighing the evidence and of granting, with sovereign power, authorisation sought by businessmen and other community leaders.

It is to be feared that the principal reasons which underly the opposition of government officials is resistance to change and the desire for power.

The Case Against Standards

Standards have many disadvantages. They are extremely cumbersome to apply and may involve overscrupulous regulations, authorisation procedures, controls, etc. Whenever standards are not respected, legal machinery must be set in motion with all the slowness and inefficiency this entails. But a pollution charge may be simpler, since the polluter pays automatically for any pollution which is emitted.

However, the main objection to standards is that they are static and provide no incentives. Standards can only be laid down after negotiations and joint studies with those concerned. For example, noise standards for motor vehicles must be based on the available technology, and who can supply this information but the manufacturers themselves? It will often be difficult to acquire knowledge of the most sophisticated technology and even more difficult to make its use compulsory for all manufacturers. To fix the limit too high would be ineffective and to fix it too low would be wasteful. It should be borne in mind that nearly all motor cars and heavy lorries conform to national and international standards, thus demonstrating that, far from acting as incentives, standards represent only the lowest common denominator. *Adoption of a standard is therefore often tantamount to endorsing the least satisfactory technique available.*

It is, of course, possible to move forward in discrete steps by updating the standards from time to time. But such progress is spasmodic and a time lag occurs, since a lot of noise is bound to have been produced by the motor vehicles before any regulations are revised, not to mention the cumbersome nature of the revision procedure. For example, it required no less than 10 years in France to reduce motor vehicle noise standards by 1 dB.

A major drawback of standards is therefore the total absence of any incentive to progress; respect of a standard is in effect a pat on the back for the polluter.

A charge would instead provide a lasting incentive to reduce noise emission until it equalled the cost of such abatement. The incentive could in fact be increased by progressively raising the rate of the charge. Charges are both economic and have an incentive effect.

They are *incentives* for the reasons given above. They are *economic* because they achieve a result at *least cost* to the community. To insist upon searching for an optimum charge rate might be an interesting theoretical exercise, but hardly one likely to lead to action. It is now time therefore to stop attempting to make the monetary evaluation of damage (social cost) the basis for an environmental policy, since it does not appear that such calculations can be made with sufficient accuracy in the foreseeable future.

On the other hand, the automatic relation between the charge and the cost of reducing noise can quite well be used as a basis. What do the economists have to say on the subject?

We can see this more easily with the aid of Fig. 6.2. The abscissa shows the level of noise reduction: noise is maximum at 0 (no reduction whatsoever) and at R_t (on the assumption, for the sake of simplicity, that it is in fact possible to eliminate the noise entirely) noise has been totally eliminated. Curve R represents the marginal cost of reducing noise, i.e. the cost of each additional unit of noise reduction. If a unitary charge is fixed at rate t, as an ordinate (that is to say, a fixed amount levied for the emission of each additional decibel—we are ignoring here the fact that, as the decibel is a logarithmic scale unit, the charge rates would have to be adjusted on that basis; for example, since an increase of three decibels doubles the amount of loudness, the tax should be doubled for each additional three dBA), the noise reduction will be fixed automatically at OR_i; beyond R_i, it will be less expensive to pay the tax than reduce the noise.

If only one polluter existed, it would be quite simple to impose a standard reduction OR_i (corresponding to maximum noise level R_i, were this the objective sought), but there are in fact a large number of polluters who have different noise reduction costs for a variety of economic and technical reasons.

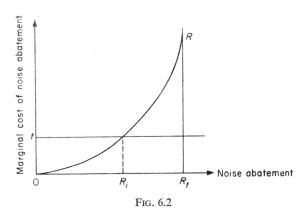

FIG. 6.2

However, if it costs A three times as much as B to achieve an equivalent reduction, it is not economic to impose the same standard on each. It is more logical to require a greater reduction from B than from A in order to achieve the greatest benefit from B's advanced technology. The total cost of the reduction by A and B will therefore be lower; in other words, the noise reduction will be more economical for the community.

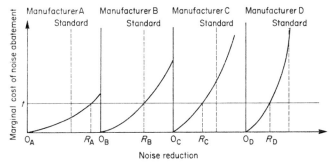

FIG. 6.3

Let us take the example of four car manufacturers: the four manufacturers shown in Fig. 6.3 have unequal marginal-cost-of-reduction curves: A, B, C and D. For a charge t, A has a substantial reduction $0_A R_A$; B a reduction $0_B R_B$; C has one of $0_C R_C$; and D has a small reduction as the marginal cost of reduction is so high. To sum up, the overall cost to the community is minimised (an economist would speak of an optimum allocation of costs by equalising the marginal costs of reduction). If a standard of R_B level had been applied generally it is clear that the total cost of reduction would have been extremely high.[6]

Such, in outline, are the main arguments for and against charges and direct regulations. What proposals can be derived from them for further action?

Problems of Implementation in the Case of Traffic Noise

While the regulatory approach has serious disadvantages, the use of charges for abating motor vehicle noise also creates difficult problems. First, a decision must be taken as to whether a charge should be levied on the motor vehicle user or the manufacturer. It must then be determined whether exclusive use of the charge, with no regulatory approach whatsoever, might not prove disastrous by producing an unduly wide range of sound levels and consequently increased annoyance. Let us examine these two points.

Who Should Pay: the Driver or the Manufacturer?

The aim of the charge system is to make the polluter pay, but circumstances vary considerably according to whether pollution comes from a stationary source (a plant) or moving source (motor vehicle). In the first case, for example, it is relatively easy to determine the noise produced by the plant and then levy the charge. In the case of motor vehicles, however, who are the polluters? They consist, in fact, of a large number of users, collectively responsible for noise. The amount of noise produced by each is impossible to determine because neither the utilisation rate of the vehicles (distance travelled), nor the driving conditions (acceleration, smoothness of flow, etc.), are known. A potentially noisy vehicle emits a varying total volume of noise, depending on whether it is driven only during holidays and at weekends or whether it is also driven daily to and from work. Furthermore, there is no common measure of the annoyance produced by the vehicle in open country or when traversing a built-up area at night. Just as the charge for water pollution will be based on the volume of pollutants discharged and the assimilative capacity of the river downstream from the discharge point, so should the noise charge depend on the amount of noise actually emitted (levels of noise emitted, length of time, at repeated intervals) and the conditions of emission (day or night, urban area, rural area, etc.).

It seems unlikely that any such measurement can be made, but even if the approximate volume of noise emitted were calculated, would a charge really be effective? The manager of a plant can react to a pollution charge by taking steps to reduce its pollution as cheaply as possible, or any other measure that maximises his profits under this new constraint (reduce production, change the production process, etc.), *since he has the economic and technical capability needed for this purpose.* But the driver of a car cannot ask a garage mechanic to reduce the noise it makes. All he can do is pay the charge, or trade in his car for a quieter model. In other words, the user has *no direct capability to reduce the intrinsic noise of his vehicle.* The only incentive provided by such a charge would be to reduce the demand for and use of noisy vehicles which had become too costly to run.

This slackening of demand would begin to have an *indirect* effect on vehicle noise by inducing manufacturers to produce quieter vehicles. Consequently, *a charge for pollution can only be effective if it directly affects an economic agent who has the economic and technical capacity required to reduce the disamenity, and the manufacturer alone has this capacity.* Accordingly, a charge levied on vehicle users is inadequate as an instrument for abating noise.

The Charge May Entail an Unduly Wide Range of Noise Levels

Would it therefore be an advantage to levy a charge on *manufacturers,* since they possess the technical and economic capability needed to reduce noise effectively? Generally speaking, we saw that an advantage of the charge is to minimise the overall cost of abatement by equalising the marginal costs. Thus, each manufacturer would respond according to his individual cost curve.

Let us assume that competing manufacturers A and B each produce a comparable range of vehicles in which each competing model has more or less equivalent noise levels. Owing to his more advanced technology, manufacturer A's marginal cost of noise reduction is lower than manufacturer B's, so he will reduce the noise of his vehicles to a lower level than that of his competitor's vehicles. Consequently, even if A and B reduce the noise level of their vehicles, the end result is likely to be less advantageous than first appears for a wider range of noise levels and a consequent greater number of peaks is obtained, and annoyance would consequently be increased. *A system of noise pollution charges is not therefore a practical proposition unless it ensures that noise levels are not only reduced but also levelled off.* Let us now consider how this problem may be solved.

A System Combining Both Standards and Charges

Although charges and direct regulations both have their disadvantages, a combination of the best features of each is quite conceivable. Since the reduction of annoyance caused by traffic noise requires that sources of noise be levelled out in a downward direction, a system of standards is needed to achieve this end. Yet such a system must remain flexible and amenable to change, which tends not to be the case with standards.

As standards achieve a minimum objective, it would likewise be extremely useful to provide a system of charges which would be levied on the user purely as an incentive. Thus, vehicle owners could very well be required to pay an annual charge based on the level of noise emitted by the vehicle's engine, which would make it more expensive to run noisy vehicles, irrespective of their other characteristics (for example, the cylinder capacity).

Drivers would therefore be encouraged to buy the least noisy vehicle in a range of very similar models, which would in turn induce the manufacturer to reduce the level of noise emitted by his model below the authorised standard in order to remain competitive. Experience has shown that vehicle taxes influence consumer choice, which in turn has an effect on the manufacturers' output. As matters now stand, taxation has a negative effect from the standpoint of noise abatement, as the heavy tax on petrol and cylinder capacity induces manufacturers to produce light cars of low or average cylinder capacity but high performance due to high compression ratio and high engine speed, which considerably increases their noisiness. A charge on noise emitted therefore also requires the reform of motor vehicle taxation.

With such a tax, which would in fact be a flat charge, manufacturers would have a standing incentive *to do better than the standard* and to seek and apply more effective noise abatement techniques. Applied in conjunction with the standard, such a charge would ensure a dynamic noise abatement strategy, one constantly capable of adjustment, while it would eliminate the disadvantage of using either system alone. Furthermore, the principle could be made more effective by increasing the rate of the charge over time.

As one result of charges is to provide funds (insofar as the rates applied are not fully effective since, if they were, noise would be reduced to the desired level and no charge would be collected), consideration might be given to setting up 'transport agencies', based on the same principle as the 'agences financieres de bassin' used for water management in France, which would be responsible for collecting such noise pollution charges and for re-investing the funds in various noise abatement projects (on a regional or local basis to be determined).

The main object is that the charges should not simply go to swell the general government Budget, *but be allocated directly to noise abatement.* We are aware that such a charge scheme is not consistent with the orthodox economic theory which states that a charge should have no direct re-distributive purposes and that no consideration should be given to the further use of the collected funds. Relative to this pure economic stand-point, we are pleading here for a 'second best' approach, i.e. a redistribu-tive charge combined with direct regulations.

Such agencies could thus help to finance roads in cuttings or noise barriers (measures to control the transmission of noise), provide grants for soundproofing certain buildings (measures to control noise at the point of reception), or help subsidise projects for re-routing certain urban highways. They might even finance some of the research and development undertaken in connection with quieting of vehicles (action at the source) or in the field of public transport.

Such agencies would achieve greater effectiveness as a result of the consistent character afforded to the system by lasting incentives and the coordination of strategy, since one and the same body would be responsible for organising all aspects of noise abatement strategy, that is to say, the incentives and aids for reducing noise at the source, participation in urban planning, grants for projects designed to reduce the propagation of noise, and the development of public transport, etc. The fund could also be allocated to help finance compensation schemes.

Furthermore, consistency would call for the collection of a motor vehicle pollution charge based on the same principle, varying according to the area, the rate rising with the level of pollution. In fact, the control of disamenities created by motor vehicles (noise, air pollution and traffic congestion) should be regarded as one and indivisible and all three aspects ought to come under the jurisdiction of the 'transport agencies'. By means of a system such as this a more effective, dynamic and coherent type of noise-abatement strategy could be achieved.

REFERENCES

1. 'A Review of Road Traffic Noise', a Report by the Working Group on Research into Road Traffic Noise, Road Research Laboratory, Report No. LR 357, United Kingdom, 1970.
2. *Chicago New Noise Regulation*, Department of Environmental Control, Chicago, 1971.
3. Simpson, R. C. *National Noise Control Policy in the United Kingdom*, OECD Urban Sector Group, April 1974.
4. *Land Compensation Act* 1973, Ch. 26, London, HMSO.
5. *Building and Builders—The Noise Insulation Regulations* 1973, Statutory Instrument No. 1363, 1973.
6. Baumol, W. J. and Oates, W. E. The use of standards and prices for protection of the environment, *Swedish Journal of Economics*, **73**, March 1971.

CHAPTER 7

THE PROSPECTS FOR REDUCING TRAFFIC NOISE

C. LAMURE

WHAT CAN BE EXPECTED FROM THE PRIVATE CAR?

Present Development of Conventional Cars

The only way to cut traffic noise levels at any given spot, whether in town, in the country, during the day or at night, is to reduce noise at its source; namely, the car itself. Any measures such as traffic engineering, or town planning, or reorganised road systems are in themselves merely palliative. However, this fact, although patently obvious, has so far led to no appreciable improvements in vehicle design. Granted that both manufacturers and administrators have had some intention of reducing noise levels, it has nevertheless to be recognised that a number of fundamental obstacles have to be overcome.

In Europe, particularly, taxation on vehicles and fuel has led to continual increases in compression ratios and, above all, in engine speeds. For example, in France from 1960 to 1970, the engine speeds of newly registered cars increased 25%—an annual increase of 2·5%. Given the relationship between engine speed and noise level stated in Chapter 4, p. 89, it follows that if European traffic volume had remained unchanged over this period, there would have been an annual increase of 0·5 dB in the average Leq.

Because the taxation system is different in the USA, America has so far escaped the consequences of this European iron rule. However, it is to be feared that increased fuel costs will eventually bring about the same situation there as in Europe. This may well bring an increased awareness of the noise problem, with resulting world-wide consequences, as with the Muskie anti-pollution legislation. For Europe, a reversal of the tendency to higher engine speeds would have produced appreciable results in five years time. In fact, giving up small engines and fitting light cars with larger cubic capacity engines has already been mentioned as one solution. The cost of such an operation would have been relatively modest, since the cost of the engine is only a small part of the total vehicle cost, and there would have been a few years delay in writing off plants where small engines are made. Unfortunately, the increased cost of fuel has somewhat dimmed the prospects for this line of action.

At a purely technical level, reduction of noise output from private cars comes up against severe constraints. Where rolling noise is not predominant, for instance, in the case of slow-moving cars or in acceleration, any fall in the present level requires action on all four noise sources which tend to share equal responsibility for generating noise. That is, the air intake, the exhaust, the engine and the fan. A reduction in the acoustic power of any one of these sources by 75% would only yield an overall reduction of 1 dB.

This would need to be followed by similar action on another source, and so on through the whole vehicle. In this way one would soon come to realise that unless engine speed can be reduced, an enormous amount of work ranging over the whole vehicle is necessary.

As a short-term measure some benefits might be hoped for from methods of avoiding excessive engine speed, such as automatic transmission. A study was carried out by the Transport Research Institute in Lyon using a Renault Model 16 saloon driven through the city streets by 12 different drivers. The noisiest driver produced a quadratic mean level, i.e. Leq when following the car, 6 dB higher than the quietest driver. However, automatic transmission tends to be costly, nor is it entirely compatible with current requirements for fuel economy. In fact, educating drivers might be a better way of reducing peak noise levels, and in any case is quite likely to result from the need to save fuel.

Rolling Noise

At cruising speeds above 40 km/h rolling noise rapidly becomes predominant. Rolling noise will probably be reduced more as a result of eliminating noisy road surfaces and tyres than from laboratory studies. For example, measurements have shown that at 50 km/h noise on a dry road might vary as much as 11 dB, according to the nature of the road and the tyres (see Chapter 4, pp. 92–94).

As road surfaces are rather long-lasting, it will take some time to get rid of the noisy ones, and at any particular place it will depend on what priority renewal of the road surface is given. The most urgent task remains the improvement of uneven road surfaces as these are usually also rather noisy.

Since tyres are replaced more frequently than road surfaces, the noisiest tyres might be eliminated in a relatively short time. However, the replacement of cross-ply by radial tyres will make very little difference since their noise levels differ by only some 2 dB at a speed of 50 km/h. The study of rolling noise has not proceeded far for automobiles, since research in Great Britain and the USA has concentrated on truck tyres.

The texture of the road surface is determined by a number of requirements. It must be cheap, non-slip, resistant to wear and tear, and drain easily when wet. It is possible to conceive of elaborately textured surfaces in urban areas, perhaps reminiscent of the days of wood-paving blocks and horse transport, derived from the carving of the surfaces to eliminate air suction between the furrows of tyre tread while avoiding excitation of the tyre carcass.

The conclusion, then, might be that the tyre tread needs further investigation. The effect of the contact surface of the tyre may lead to attempts to reduce noise here, though without reducing tyre adhesion in wet weather. Yet an overall tyre–road equilibrium must be established, for the cost of high-quality road surfaces, even in limited areas, has to be compared with that of renewing tyres.

Innovations—What Will Be the Typical Urban Car?

Though great hopes have been pinned on electric cars with rechargeable batteries, these have been somewhat dashed by the limited range (60–100 km) and the modest speeds (60 km/h) achieved by such vehicles. Again, prices would appear to be double those of conventional cars and the problem of recharging is a serious one, immobilising the vehicle for long periods of time. Moreover, such cars will still emit noise, even if only rolling noise. Measurements made on a 4 hp standard model fitted with an electric battery yielded the following results in dBA:

Speed (km/h)	20	30	40		50		60	70	
Gear ratio	in 2nd	in 2nd	in 2nd	3rd	in 2nd	3rd	in 3rd	in 3rd	4th
Petrol engine	53·5	61·5	67	61	71	65	68	71	70
Electric vehicle	50	55	58·5		61·5		63·5		
Difference	−3·5	−6·5	−8·5	−2·5	−9·5	−3·5	−4·5		

It might well be asked whether conventional petrol-driven vehicles could not be designed to function at similar noise levels if restricted to similar speeds and performance, as has already been noted in the case of buses (Chapter 5, p. 133), and the same remarks apply to 'hybrid vehicles' using heat engines at constant speed together with accumulators. Such 'hybrids' have, in fact, been considered as an anti-pollution measure (see Chapter 5, p. 136).

New, non-explosive heat engines, such as the Rankine Cycle Steam or the Stirling engine may arouse expectations, although at present they can only function effectively in sizes above 200 CV. On the whole, nothing very novel has appeared which seems likely to lead to silent vehicles with performance comparable to the present European car.

By prohibiting conventional car traffic in large central areas it might be possible to introduce small, lightly built vehicles with low rates of acceleration and safety standards hardly possible in ordinary streets. Such vehicles, particularly electrically driven ones, have been produced by various concerns. Is it possible to imagine that the success of 'pedestrian only' precincts presages the future success of similar precincts for specifically urban cars?

But will such vehicles compete effectively with public transport or with pedal cycles? In any case, they provide no solution to transport in the outlying suburbs.

The Slow Replacement of Present-Day Cars

Assuming that quieter vehicles selling at competitive prices were introduced, nevertheless their effect on the noise level would take a long time to become apparent. Table 7.1 gives the reduction in peak level, L_m and Leq resulting from the introduction of vehicles which, at speeds similar to those of conventional vehicles, would have noise levels 5, 10 and 15 dBA lower. It is noteworthy that if the new vehicles were 10 dBA quieter—which would itself be quite an achievement—the effects would only be appreciable if not more than 10% of total traffic were made up of conventional vehicles. As for noise measured by average level and LNP, the decrease would be even smaller or there might even be an increase in certain circumstances. Even with the present rather high vehicle replacement rate, a fall of 8 dBA in Leq level will only be achieved by 1985, so that extremely quiet vehicles, i.e. 20 dBA reduction, could not have any significant effect by comparison with 10 dBA reduction vehicles before then (Fig. 7.1). Now the energy shortfall and the fact that quiet vehicles will be more expensive will certainly lower the replacement rate, so that when estimating the time needed to improve the general situation, at least another 5 years must be added to launch the new vehicles.

On the whole, therefore, what may be hoped for from improvements in cars has to come down to something relatively modest. We may simply look

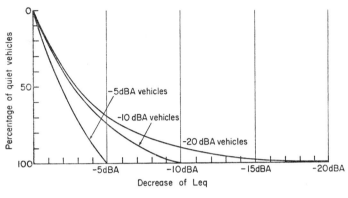

FIG. 7.1

for a decrease of 5 dBA in Leq over the next 10–15 years, assuming no further increase in traffic volumes, which would be a rather novel development (Table 7.1). Indeed, reducing the overall volume of traffic by 10–20%

TABLE 7.1

Quiet vehicles with emission level lower by	−5 dBA		−10 dBA		−15 dBA	
	L_m	Leq	L_m	Leq	L_m	Leq
Percentage of quiet vehicles in total traffic						
10	0	0	0	0	0	−1
50	0	−2	0	−2·5	0	−3
90	0	−4	0	−8	0	−9
100	−5	−5	−10	−10	−15	−15

could have more effect than substituting cars 5 dB quieter in the existing traffic volume, for it would be equivalent to replacing them with vehicles 15 dB quieter.

WHAT CAN BE EXPECTED FROM COMMERCIAL VEHICLES?

As soon as the percentage of commercial vehicles exceeds 10% of the traffic volume, their general contribution to the Leq level is equivalent to 10 cars for every truck, and works out even larger for L_{max}, LNP, TNI, and consequently annoyance at night. Thus total replacement of present trucks by new models about 10 dB quieter may, in effect, be compared to replacement of 10 present-day cars by models 10 dB quieter. We shall see that this is within present industrial and economic possibilities.

The decrease in sound levels may be measured by two indices, one estimating the average and the other the fluctuating noise level. Employing L_{10} and TNI, Nelson and Fanstone[1] calculated the reduction to be expected with different proportions of trucks in various volumes of total

TABLE 7.2

REDUCTION IN L_{10} AND TNI FOR VARIOUS TRAFFIC FLOWS
AND PERCENTAGES OF LORRIES[1]

(*a*) Lorries reduced by 10 dBA

	200 *veh/h*		600 *veh/h*		2 000 *veh/h*	
	ΔL_{10}	ΔTNI	ΔL_{10}	ΔTNI	ΔL_{10}	ΔTNI
5% lorries	0·7	1	0·7	1	1·1	2
20% lorries	2·5	2	2·9	3	4·2	9
40% lorries	4·5	3	5·8	8	6·3	13

(*b*) Cars reduced by 5 dBA

	200 *veh/h*		600 *veh/h*		2 000 *veh/h*	
	ΔL_{10}	ΔTNI	ΔL_{10}	ΔTNI	ΔL_{10}	ΔTNI
5% lorries	4·5	5	4·5	5	3·5	1
20% lorries	3·5	5	2·1	0	0·5	−7
40% lorries	2·1	3	0·3	−4	0·1	−5

(*c*) Lorries reduced by 10 dBA and cars reduced by 5 dBA

	200 *veh/h*		600 *veh/h*		2 000 *veh/h*	
	ΔL_{10}	ΔTNI	ΔL_{10}	ΔTNI	ΔL_{10}	ΔTNI
5% lorries	5·4	5	5·4	5	5·8	7
20% lorries	6·3	5	6·7	7	7·8	12
40% lorries	7·3	6	8·6	11	9·1	14

traffic, using a TRRL computer program. Their data have been grouped in Table 7.2. It can be seen that where traffic is heavy or includes a high proportion of heavy vehicles, improvements restricted to private cars yield poor results, even increasing values of TNI and aggravating the situation. On the other hand, for the same conditions, confining improvements to heavy vehicles is extremely effective and it is therefore very reasonable to direct vehicle improvements initially to heavy lorries.

Programme for Construction of Demonstration Quiet Heavy Vehicles

The technical and economic possibilities for quieting heavy lorries are better than those for private cars. As with buses, there is more space and load capacity available for silencers, resonators and engine insulation while, unlike buses, trucks with engines over 200 hp increasingly use super-charging. Apart from the basic advantage of improved coupling power to weight ratio, supercharging flattens the curve of rising cylinder pressure following ignition, while the supercharger turbine lies in the exhaust flow. For these reasons both engine and exhaust noise are reduced and the overall level is cut by about 6 dBA.

In the UK a research programme to evolve two types of articulated heavy truck is in progress. These will yield noise levels at least 10 dBA lower than those of present vehicles.[2] This programme envisages three years' research and a further two years of development. Work on the design, production and testing of the engine, air intake and gearbox is being carried out by the Institute of Sound and Vibration at the University of Southampton, while the Transport and Road Research Laboratory is responsible for new axle design and reduction of tyre noise. The Motor Industry Research Association is working on the design, production and testing of the cab, the cooling and the exhaust system. Targets for maximum noise levels (in dBA) for the various components are listed below:

		At 7·5 m
ISVR	Engine, including gearbox	77
	Air intake	69
TRRL	Tyre–road surface	75–77
	Rear axle	69
MIRA	Cooling system	69
	Exhaust system	69

In the US three programmes have been initiated by the Department of Transportation, with manufacturers and acoustic consultants working together.

For one programme the general contractor is United Freightlines, with BBN and others as subcontractors.[3] They have worked on a six-cylinder turbocharged 350 hp truck in three ways: (a) they enclosed engine and transmission with an aluminium sheet metal box lined with glassfibre padding 2 in thick; (b) they reduced the speed of the fans or stopped them entirely when not needed; (c) they employed a more efficient exhaust system.

The results are interesting: the maximum acceleration noise is reduced by 11 dBA from 88 to 77 dBA at a distance of 50 ft (testing procedure is SAE 366a). The quieter truck is 10 dBA quieter at 30 mph and 6·5 dBA quieter at 50 mph. The relatively higher back pressure is offset against the relatively lower radiator fan drive requirement. Fuel consumption was 9% less for the same journey compared to normal performance, because on the standard truck the fans require 16–21 hp.

A kit suitable for any truck has been designed, costing about $1600. No basic research on the engine or silencer has been necessary; enclosing and cooling the engine and decoupling the fan are the main methods employed.

In France, Berliet and Saviem received government grants to build trucks with sound levels at least 8 dBA below the levels emitted by current lorries. In Germany and in Sweden (Volvo), industrial research has been successful, while turbocharging has enabled Volvo to build trucks only a few dBA noisier than private cars.

Prospects for Quiet Trucks

Thus, the present objectives of American and British plans for 'quiet trucks' are a reduction of ISO noise levels of about 10 dBA, and it seems perfectly possible to achieve this in two years' time.

In 10 or 15 years' time we should see trucks driven by steam or Stirling engines. Their development in the USA (Chrysler), in Germany (MAN), in Holland (Philips) and in Sweden (United Stirling), may be speeded up by the energy crisis, for they can use more varied fuels than the diesel engine and are completely non-polluting.

On the whole, a notable improvement of trucks is already going on; in five years' time it should lead to a good proportion of trucks emitting hardly any more noise than cars. In particular, noise reduction in LNP will be quite considerable. However, taking account of the rather variable replacement rates of truck fleets, progress will depend on the general prosperity of road transport enterprises, on the preferences for road or rail transport, and of course on the regulation of noise levels by different countries. European countries do not seem to be in the lead at present, but the success of American and British efforts may hasten the trend to lower the limits from 90 to about 77–80 dBA in terms of the ISO norm.

Freight Transport Depots

The effects of organising freight storage depots and deliveries on a rational system in major towns, as mentioned in Chapter 5, pp. 161–162, would be rather limited in practice, since future quieted trucks and delivery vans would all tend to produce only moderate levels of noise. Some improvements might on the other hand be expected by transferring more freight from road to rail.

DEVELOPMENT OF NOISE DISPERSION IN TIME AND SPACE

The increase of average noise levels is particularly important in outlying areas or in small towns, since in these areas the annual increase of private car traffic is about 3 %, that is, twice as much as in central areas. In addition, this traffic extends in every direction at night, particularly around the main arterial roads for freight transport.

The following developments may be considered in the light of what has been said in Chapter 5:

1. In the central parts of cities the background noise in 5 or 15 years time will depend above all on:
   (a) the restriction of the number of conventional vehicles in the traffic;
   (b) the possible introduction, by then, of pedal cycles or specially prepared vehicles.
   (c) the improvement of public transport.

This last may indeed profit most from the large range of technical possibilities, and also from the energy shortfall. The potential advantages of modern diesel buses would ensure rapid developments which, in practice, depend only on town council decisions.

2. In major outlying suburban areas the expansion of public transport will hardly cause the conventional automobile to disappear, except in a few high-density housing projects. Improvement will come particularly from quieter trucks or delivery vans, but this will only benefit people living near major arterial roads.

As regards future building design, there may be advantages to be gained from new layout plans using cul-de-sacs and courtyards.

3. People living near heavy transport routes will perhaps benefit mainly from the improvement of heavy vehicles and from a shift towards other transport media which use less energy. It should also be possible to lengthen the comparatively quiet period of the night up to 8 hours.
4. People living in small towns and villages, or in detached houses on an estate, will probably experience little change.

Over the next 15 years it is only possible to conceive of limiting the spread of noise throughout the length and breadth of the land if the volume of traffic is itself reduced. Given the growing energy problem, this may well take place.

REFERENCES

1. Nelson, P. M. and Fanstone, J. 'Estimates of the Reduction of Traffic Noise following the Introduction of Quieter Vehicles', TRRL Laboratory Report 624, 1974.
2. Watkins, L. H. A quiet heavy lorry, *Commerical Motor*, March 22, 1974.
3. Kaye, M. C. and Ungar, E. E. 'Acoustic and Performance Test Comparison of Initial Quieted Truck with Contemporary Production Trucks', DOT TST, 74.2, Sept. 1973.

INDEX

Index 219